Advanced and Intelligent Manufacturing in China

Series Editor

Jie Chen, Tongji University, Shanghai, Shanghai, China

This is a set of high-level and original academic monographs. This series focuses on the two fields of intelligent manufacturing and equipment, control and information technology, covering a range of core technologies such as Internet of Things, big data, 3D printing, robotics, intelligent equipment, industrial network security, and artificial intelligence, and epitomizing the achievements of technological development in China's manufacturing sector.

With Prof. Jie Chen, a member of the Chinese Academy of Engineering and a control engineering expert in China, as the Editorial in Chief, this series is organized and written by more than 80 young experts and scholars from more than 40 universities and institutes. It typically embodies the technological development achievements of China's manufacturing industry. It will promote the research and development and innovation of advanced intelligent manufacturing technologies, and promote the technological transformation and upgrading of the equipment manufacturing industry.

Tongying Guo · Hui Zhang · Lincang Zhu

Special Robot Technology

Chemical Industry Press Co., Ltd.

Springer

Tongying Guo
Shenyang Jianzhu University
Shenyang, Liaoning, China

Hui Zhang
Shenyang Jianzhu University
Shenyang, Liaoning, China

Lincang Zhu
JD Zhilian Cloud Platform
Beijing, China

B&R Book Program

ISSN 2731-5983 ISSN 2731-5991 (electronic)
Advanced and Intelligent Manufacturing in China
ISBN 978-981-99-0588-1 ISBN 978-981-99-0589-8 (eBook)
https://doi.org/10.1007/978-981-99-0589-8

Jointly published with Chemical Industry Press
The print edition is not for sale in China (Mainland). Customers from China (Mainland) please order the print book from: Chemical Industry Press.

This Springer imprint is published by the registered company Springer Nature Singapore Pte Ltd.
The registered company address is: 152 Beach Road, #21-01/04 Gateway East, Singapore 189721, Singapore

About This Book

This book systematically introduces the basics of special robots, path planning algorithms, and application examples of ruin search and rescue robots and text questions and answers robots. Most of the content covers hot issues of research in the field of robotics in recent years and is an accumulation and summary of the authors' research results in the field over the years.

The book is divided into seven chapters. The main contents are the definition and classification of robots, the development status and core technologies of special robots, the main application areas, the drive system, mechanism and sensing technology of robots, the positioning algorithm and path planning algorithm of mobile robots, the system composition and autonomous motion control research of ruin search and rescue robots, and the architecture, key technologies, and typical applications of text questions and answers robots.

This book can be used as a reference book for scientific researchers and engineers engaged in the research and development of special robots and their applications, or as a textbook for graduate students or senior undergraduates in control science and engineering, computer science and technology, mechanical and electronic engineering, etc.

Contents

Chapter 1
Introduction

1.1 Definition and Classification of Robots

1.1.1 Definition of Robot

As one of the greatest inventions of mankind in the twentieth century, robotics has played an important role in both manufacturing and non-manufacturing fields. With the rapid development of robotics and the advent of the information age, new types of robots are emerging, the content covered by robots is getting more abundant, and the definition of robots is being enriched and innovated.

The word "robot" first appeared in the 1920 science fiction script "Rossum's Universal Robots" by Czech writer Karel Capek. In the script, The Czech word "Robota" was written into "Robot" by Capek, which means forced labor. The play, which foreshadowed the tragic impact of the development of robots on human society, attracted widespread attention and is credited as the origin of the word robot. Kepek raised the issues of safety, perception and self-reproduction of robots. Advances in science and technology are likely to raise problems that humans do not want to have. Although the world of science fiction is only an imaginary one, human society will likely face this reality.

To prevent robots from harming humans, in 1950 science fiction writer Isaac Asimov proposed the "Three Principles of Robotics" in his book 《I, Robot》.

- The robot must not harm humans, nor be allowed to stand by and do nothing when it sees that humans will be harmed.
- Robots must obey human orders unless the human orders violate the above first article.
- The robot must protect itself from harm, unless this is contrary to the above two articles.

These three principles gave a new ethical dimension to the robotics society. To this day, it will continue to provide very meaningful guidelines for robotics researchers, design manufacturers, and users.

T. Guo et al., *Special Robot Technology*, Advanced and Intelligent Manufacturing in China, https://doi.org/10.1007/978-981-99-0589-8_1

At the first academic conference on robotics held in Japan in 1967, two representative definitions were proposed. One is proposed by Masahiro Mori and Chohei Hopeda, that is a robot is a flexible machine with seven characteristics such as mobility, individuality, intelligence, versatility, semi-mechanical and semi-humanity, automaticity, and slavery. From this definition, Masahiro Mori further proposed to express the image of robot with 10 characteristics such as automaticity, intelligence, individuality, semi-mechanical and semi-humanity, operability, versatility, information, flexibility, finiteness, and mobility, etc. The other one is proposed by Ichiro Kato that a machine with the following three conditions can be called a robot.

- Individuals with three elements: brain, hands and feet.
- Having non-contact sensors (receiving distant information with the eyes and ears) and contact sensors.
- Sensors with balance and positioning.

The definition emphasizes that a robot should have humanoid characteristics, i.e., it works by hand, moves by foot, and the brain performs the task of unified command. Non-contact and contact sensors are equivalent to the five human senses, enabling the robot to recognize the external environment, while balance and positioning are indispensable sensors for the robot to perceive its own state.

The Robotics Institute of America (RIA) defines a robot as a reprogrammable, multifunctional operator used to move a variety of materials, parts, tools or other special devices.

The Japanese Industrial Robot Association (JIRA) defines a robot as a general-purpose machine with memory and end-effectors that can replace human labor through automated actions.

The International Organization for Standardization (ISO) defines a robot as a machine that can be programmed and automatically controlled to perform tasks such as working or moving.

Our scientists define a robot as an automated machine, with the difference that such a machine has some intelligent capabilities similar to those of a human or a living being, such as perception, planning, movement and synergy, and is an automated machine with a high degree of flexibility.

With the growing awareness of the intelligent essence of robotics, robotics has begun to penetrate into various fields of human activity. With the application characteristics of these fields, a wide variety of special robots and various intelligent robots with sensing, decision-making, action and interaction capabilities have been developed. Although there is no strict and precise definition of robot, we hope to grasp the essence of robot: a robot is a machine device that performs work automatically. It can either accept human command, run pre-written programs, or act according to principles developed by artificial intelligence technology. Its task is to assist or replace the work of humans. It is the product of advanced integration of cybernetics, mechatronics, computers, materials and bionics. It has important uses in industry, medicine, agriculture, service industry, construction and even military.

1.1.2 *Classification of Robots*

There is no unified standard for the classification of robots internationally, and there are different classifications from different perspectives.

- Classification from the perspective of application environment

Currently, robotics experts in China classify robots into two major categories from the application environment, namely industrial robots and special robots. International robotics scholars, from the application environment, also divide robots into two categories: industrial robots in manufacturing environment and service in non-manufacturing environment and humanoid robots, which is consistent with the classification in China.

Industrial robots are multi-joint manipulators or multi-degree-of-freedom robots for industrial applications. Special robots are various kinds of advanced robots other than industrial robots that are used in non-manufacturing industries and serve human beings, including: service robots, underwater robots, entertainment robots, military robots, agricultural robots, medical robots, etc. Among the special robots, some branches are developing rapidly and have a tendency to become independent systems, such as service robots, underwater robots, military robots, micro-operated robots, etc.

- Classification by control mode
  - Operated robots: Capable of automatic control, reproducible programming, multi-functional, with several degrees of freedom, fixed or movable, used in related automation systems.
  - Program-controlled robot: The mechanical actions of the robot are controlled sequentially according to the pre-required sequence and conditions.
  - Teach-in/Playback robot: Through guidance or other means, the robot is first taught to move and input the work program, and then the robot automatically repeats the work.
  - CNC robot: It is not necessary to make the robot move, but to teach the robot by numerical values and language, and the robot operates according to the information after the teaching.
  - Sensory-controlled robots: The information obtained by sensors is used to control the robot's movements.
  - Adaptation-controlled robots: Robots can adapt to changes in the environment and control their own actions.
  - Learning-controlled robot: The robot can "experience" the work experience, has certain learning functions, and use the "learned" experience in the work.
  - Intelligent robot: At least three elements, one is the sensory elements, used to recognize the state of the surrounding environment; second is the motor elements, according to the external world to make reactive action; third is the thinking element, according to the information obtained by the sensory elements, to determine what kind of action to use.

- Classification by robot mobility

They can be divided into semi-mobile robots (where the robot is fixed in a certain position as a whole and only some parts can move, such as robots) and mobile robots.

With the continuous development of robotics, it has been found that robots operating at a fixed location do not fully meet the needs of all aspects. Therefore, in the late 1980s, many countries systematically carried out research on mobile robotics. The so-called mobile robot is a robot with a high degree of autonomous planning, self-organization, and self-adaptive capabilities, suitable for working in complex unstructured environments, which incorporates computer technology, information technology, communication technology, microelectronics, and robotics. Mobile robots have mobile functions, and have greater mobility and flexibility than general robots in replacing people in dangerous and harsh (e.g., radiation, toxic, etc.) environments and in environments where people are not available (e.g., cosmic space, underwater, etc.).

Mobile robots can be classified from different perspectives. According to the robot's movement method, it can be divided into wheeled mobile robots, walking mobile robots (single-legged, double-legged and multi-legged), tracked mobile robots, crawling robots, creeping robots and swimming robots, and other types; according to the working environment, it can be divided into indoor mobile robots and outdoor mobile robots.

1.2 Development History and Trends of Robots

Robot is an automation equipment that integrates advanced technologies in multiple disciplines such as mechanics, electronics, control, sensing and artificial intelligence. The application of robotic systems can not only help people get rid of some dangerous, harsh, hard-to-reach and other environmental operations (such as dangerous object demolition, mine clearance, space exploration, undersea exploration, etc.), but also reduce people's labor intensity, increase labor productivity and improve product quality because robots have the characteristics of high operating accuracy and inexhaustibility.

Since the birth of the world's first robot, robotics has developed rapidly. The application of robots has also expanded from industrial manufacturing to many fields such as military, aerospace, service industry, medical care, and human daily life. The integration and development of robotics with artificial intelligence technology, advanced manufacturing technology and mobile Internet technology has driven changes in the way of life of human society. The robotics industry is also gradually becoming a new high-tech industry.

1.2.1 Development History of Robots

The history of robot development can be divided into four stages, as shown in Fig. 1.1.

The first stage, the budding development period. In 1954, George DeVore, an American, got a patent for building the world's first programmable robot. This robot was capable of performing different jobs according to different programs, and therefore had versatility and flexibility. In 1959, DeVore teamed up with American inventor Joseph Ingeborg to build the first industrial robot. This was followed by the establishment of Unimation, which is the world's first robot manufacturing facility. In 1956, at the Dartmouth Conference, Marvin Minsky presented his view of intelligent machines: "they can create abstract models of their surroundings, and if they encounter a problem, they can find a solution from the abstract model." This definition influenced the direction of intelligent robotics research for the next 30 years. During this period, robots entered a practical phase with the development of mechanism and servo theory.

The second stage, industry gestation period. In 1962, American Machine and Foundry (AMF) built the world's first cylindrical coordinate robot, named Verstran, which means "universal moving", and successfully used it in the Ford Motor Company in Canton, USA. In 1969, Japan developed the first robot that walked with two arms. At the same time, Japan, Germany and other countries faced labor shortage and other problems, so they invested heavily in research and development of robots, and the technology developed rapidly to become a robot powerhouse. At this stage, with the development of computer technology, modern control technology, sensing technology, and artificial intelligence technology, robots also developed rapidly. The robots in this period were "Teach-in/Playback" robots, which only had the ability to memorize and store, repeat the work according to the corresponding procedures,

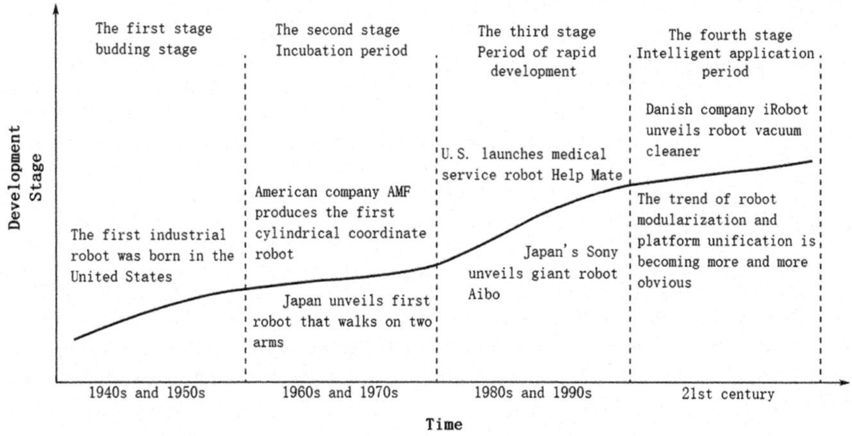

Fig. 1.1 Robot development history

and basically had no perception and feedback control ability to the surrounding environment.

The third stage, rapid development period. 1984, the United States launched medical service robots (Help Mate), which can deliver meals, medicine and mail to patients in hospitals. 1999, Sony Japan launched large robots Aibo. At this stage, sensory robots emerged with the development of sensing technologies, including visual sensors, non-visual sensors (force, touch, proximity, etc.), and information processing technologies. Robots for welding, painting, and handling were widely used in industrial industries. iRobot of Denmark introduced dust-vacuuming robots in 2002. Currently, vacuuming robots are the largest selling household robots in the world. since 2006, the trend toward modularization of robots and unification of platforms has become increasingly evident. In the past five years, the average annual growth rate of global industrial robot sales has exceeded 17%. Meanwhile, service robots have developed rapidly and their applications have become increasingly widespread, with medical rehabilitation robots represented by surgical robots forming a larger industrial scale, and special operation robots such as space robots, bionic robots and disaster rescue robots realizing applications.

The fourth stage, intelligent application period. Since the beginning of the twenty-first century, with the continuous improvement of labor costs and technological progress, countries have successively carried out the transformation and upgrading of manufacturing industries, and there has been a boom in the replacement of people by robots. At this stage, with the iterative upgrading of perception, computing, control and other technologies and the in-depth application of artificial intelligence technologies such as image recognition, natural speech processing and deep cognitive learning in the field of robotics, the trend of servitization in the field of robotics has become increasingly obvious, gradually penetrating into every corner of social production and life, and the scale of robotics industry has also grown rapidly.

1.2.2 Development Trend of Robots

Since entering the twenty-first century, intelligent robotics has been rapidly developing, with the following specific development trends.

- Rapid development of sensing technology

Sensing technology, which is the foundation of robotics, has seen new developments and various new sensors have emerged, such as ultrasonic tactile sensors, electrostatic capacitive distance sensors, 3D motion sensors based on fiber optic gyroscopic inertial measurements, and vision systems with workpiece detection, identification and positioning functions. Multi-sensor integration and fusion technologies are gaining application on intelligent robots. Single sensing signal is difficult to ensure the accuracy and reliability of input information, and cannot meet the requirements of intelligent robot systems to obtain comprehensive and accurate environmental information to enhance decision-making capabilities. Using multi-sensor integration and fusion technology, sensing information can be used to obtain a correct understanding of the environment, making the robot system fault-tolerant and ensuring fast and correct information processing of the system.

- Using modular design technology

The structure of intelligent robots and advanced industrial robots should strive to be simple and compact, and the design of their high-performance components, or even all mechanisms, has developed in the direction of modularity; their drives use AC servo motors and develop in the direction of small size and high output; their control devices develop in the direction of miniaturization and intelligence, using high-speed CPUs and 32-bit chips, multiprocessors and multifunctional operating systems to improve the real-time and rapid response of robots. The modularity of the robot software simplifies programming. The modularization of robot software simplifies programming, develops off-line programming techniques, and improves the adaptability of robot control systems.

The application of robots in production engineering systems allows automation to evolve into integrated flexible automation and the intelligent and roboticization of production processes. In recent years, robotic production engineering systems have gained continuous development. The automotive industry, engineering machinery, construction, electronics and motor industries, and the home appliance industry are introducing advanced robotics in the development of new products, adopting flexible automation and intelligent equipment, and transforming existing means of production, resulting in an upward trend in the development of robots and their production systems.

- Breakthrough in micro-robot development

Micro-machines and micro-robots are one of the cutting-edge technologies of the twenty-first century. We have already developed finger-sized miniature mobile robots that can be used to enter small pipes for inspection operations. It is expected that millimeter-sized micro mobile robots and medical robots with a diameter of a few hundred microns will be produced, which will allow them to enter directly into human organs for diagnosis and treatment of various diseases without harming human health. Micro actuators are one of the basic and key technologies for the development of micro robots. It will have a significant impact on precision machining, modern optical instruments, ultra-large scale integrated circuits, modern bioengineering, genetic engineering and medical engineering. Micro-robots will be of great use in the above-mentioned engineering.

- Breakthrough in the development of new robots

Telepresence or telexistence, is known as proxemics. This technique is capable of measuring and estimating the anthropomorphic motion and biological state of a person to a predicted target, displaying field information for designing and controlling the motion of anthropomorphic mechanisms.

Virtual reality (VR) technology is a newly researched intelligent technology that recombines the reality of events after decomposing them in time and space. This technology includes three-dimensional computer graphics technology, multi-sensor interactive interface technology, and high-definition display technology. Virtual reality technology can be applied to areas such as remote-controlled robotics and proximity communication. For example, a Mars exploration robot could be operated remotely from Earth to collect soil on the Martian surface.

Shape-memory alloys (SMA) are known as smart materials, and the resistance of SMA changes with temperature, causing the alloy to deform and be used to perform actuation actions and perform sensing and driving functions. Reversible shape memory alloys (RSMA) have also been used in micro-robots.

Multiple autonomous robotic systems (MARS) is another intelligent technology that has been explored in recent years, and it emerged from the development of single intelligent machines that require coordinated operations. Multiple robots have a common goal to accomplish interrelated actions or operations, and MARS is a development of swarm-controlled robotic systems with consistent operational goals, shared information resources, and individual subjects of local (decentralized) actions sensing, acting, being controlled, and coordinating in a global environment. Among many new intelligent technologies, the development and application of artificial neural network-based recognition, detection, control, and planning methods occupy an important position. Robot planning based on expert systems has gained new developments and is used for automatic grasping in addition to task planning, assembly planning, handling planning, and path planning.

- Gradual improvement in the autonomy of mobile robots

In recent years, research on mobile robots has received attention, and autonomous mobile robots are one of the most researched ones. Autonomous robots can make global path planning based on known map information in accordance with pre-given task instructions, and constantly sense the local environment information around them during the travel process, make decisions autonomously, guide themselves around obstacles, drive safely to the specified target, and perform the required actions and operations. Mobile robots have a wide range of application prospects in industry and national defense, such as cleaning robots, service robots, patrol robots, defense and chemical reconnaissance robots, underwater autonomous operation robots, flying robots, etc.

- Language communication function is becoming more and more perfect

The language function of modern intelligent robots relies on the pre-storage of a large number of speech utterances and textual vocabulary utterances in their internal memory, and their language capability depends on the size of the amount of utterances stored in the database to the range of languages they store. Obviously, the larger the database vocabulary is, the more capable the robot will be in chatting. We can further assume that if a bot has a large enough database to cover all the words and phrases, then it is possible for the bot to be comparable to, or even better than, the chatting ability of a normal human. At this point, the bot would have a broader knowledge base, although the bot may not be aware of the true meaning of the chat phrases.

In addition, the robot needs to have the ability to reorganize its own language vocabulary. When a human communicates with it, if it encounters a statement or vocabulary that is not in the language packet program, it can automatically respond with a new sentence by using related or similar words and phrases, according to the structure of the sentence, which is also equivalent to human learning and logic capabilities, a kind of consciousness.

- Perfection of various movements

The movements of robots are relative to the imitation of human movements, and we know that the movements that humans can do are extremely diverse. But modern intelligent robots can also imitate some of the movements of people, but relatively is a bit rigid feeling, or the movement is relatively slow. Future robots will be more flexible human-like joints and simulated artificial muscles, making their movements more human-like, imitating all the movements of people, and even do more tangible will become possible. It is also possible to make some movements that are difficult for ordinary people to make, such as somersault, handstand, etc.

- The ability of logic analysis is getting stronger and stronger

In order to imitate human perfectly, scientists will keep giving it many logic analysis program functions in the future, which is also equivalent to the performance of intelligence. For example, the ability to reorganize words into new sentences is a perfect expression of logic, and if it is low on energy, it can recharge itself without the help of its master, which is a kind of consciousness. In short, logical analysis helps robots to do many tasks on their own, without the help of humans, and to help them as much as possible to accomplish some tasks, even more complex ones. At a certain level, robots with strong logical analysis capabilities have more advantages than disadvantages.

- Diverse functions with increasingly

The purpose of human-made robots is to serve humans, so it will become as multi-functional as possible, for example, in the family. It can sweep the floor, vacuum, and be your friend to talk about the day, and it can watch your children. When you go outside, the robot can help you move some heavy things, or lift some things, and even as your personal bodyguard. In addition, the future of advanced intelligent robots will also have a variety of deformation functions, such as from a humanoid state, into a luxury car, it carries you to where you want to go, this more ideal vision, in the future are possible to achieve.

- More human-like in appearance

Scientists develop more and more advanced intelligent robots, is mainly in the human form as a reference object. Naturally first need to have a very simulated human appearance is the primary premise, in this regard Japan should be relatively leading, Our country is also very good.

For future robots, the degree of simulation is likely to reach even if you look at its appearance close at hand, you will only take it as a human, it is difficult to distinguish is a robot, this situation is like the American science fiction blockbuster "Terminator" in the machine character shape has the most perfect human appearance.

1.3 Development Status and Core Technologies of Special Robots

Special robots are various advanced robots used in non-manufacturing industries and serving human beings, in addition to industrial robots, and have always been the focus of intelligent robotics research. Compared with industrial robots in the manufacturing industry, special robots in the non-manufacturing field are characterized by the unstructured and uncertain working environment, and thus have higher requirements on robots, which require robots with walking function, external sensing capability and local autonomous planning capability, which is an important development direction of robotics.

1.3.1 Global Special Robots Development Status

In recent years, the performance of global special robots has continued to improve, giving rise to emerging markets and attracting high attention from governments. In 2017, the global market size of special robots has reached \$5.6 billion, and according to the average annual growth rate of 12%, the global market size of special robots will reach \$7.7 billion by 2020 (see Fig. 1.2). The U.S., Japan, and the EU are the global leaders in innovation and marketing of specialty robots. The U.S. has proposed a "Robotics Roadmap" with special robots as the key development direction for the next 15 years. Japan has proposed a "robot revolution" strategy, covering three main directions: special robots, industrial robots in the new century, and service robots, and plans to double the market size to 12 trillion yen by 2020, with special robots being the fastest growing area. The European Union has launched the world's largest civilian robotics research and development project and plans to invest 2.8 billion euros by 2020 to develop robotic products, including special robots, and quickly bring them to market.

Currently, special robots are developed with the following characteristics.

- Technological advances promote increase in the level of intelligence.

The traditional programmed and remote-controlled robots are difficult to respond effectively when the environment changes rapidly due to fixed procedures and

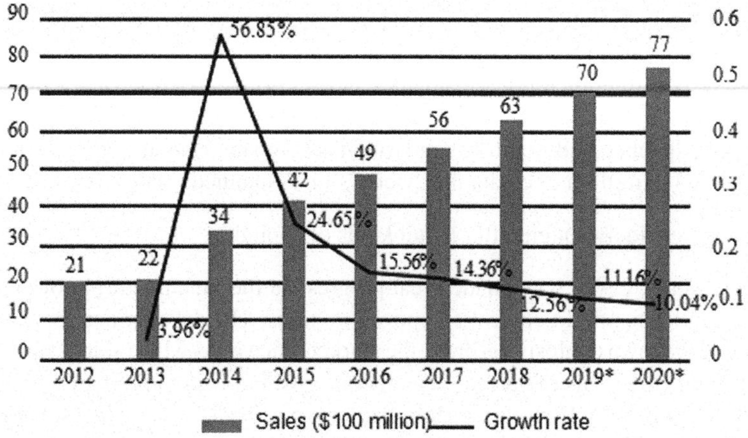

Fig. 1.2 Global specialty robotics sales and growth rate 2012–2020

long response times. With the continuous progress of sensing technology, bionic and biological modeling technology, and biomechanical information processing and identification technology, special robots have gradually realized the closed-loop workflow of "sensing-decision-behavior-feedback" and have preliminary autonomous intelligence. At the same time, the new bionic materials and rigid-flexible coupling structure have further broken the traditional mechanical mode and improved the environmental adaptability of special robots.

- Replace humans in more special environments to engage in dangerous labor.

Current special robots have a certain level of autonomous intelligence, through the integrated use of vision, pressure and other sensors, deep integration of hard and soft systems, as well as continuous optimization of control algorithms, special robots have been able to complete the positioning, navigation, obstacle avoidance, tracking, QR code recognition, scene perception recognition, behavior prediction and other tasks. For example, Boston Dynamics has released Handle, a two-wheeled robot that achieves stable control of jumping while gliding rapidly. With the increasing intelligence and adaptability to the environment of special robots, they have very broad application prospects in many fields such as military, riot control, firefighting, extraction, construction, transportation, security monitoring, space exploration and pipeline construction.

- Disaster relief, bionic and manned fields have gained high attention.

In recent years, natural disasters, terrorist activities, and armed conflicts around the world have posed a great threat to the safety of people's lives and property. In order to improve crisis response capabilities, reduce unnecessary casualties as well as fight for the best rescue time, governments and related institutions have invested heavily to increase support for the research and development of disaster relief, bionic, manned, and other special robots, such as the Japanese researchers who developed a disaster relief robot based on the created a remotely controllable two-armed disaster search and rescue construction robot. Meanwhile, Japan's SoftBank Group has acquired two bionic robotics companies, Boston Dynamics and Schaft, owned by Google's parent company Alpahbet, and South Korean robotics company Hantai Future Technology has spent $216 million to create the "world's first" manned robot.

- UAVs are widely sought after by all kinds of capital.

In recent years, UAVs have made great progress in the manufacturing of complete aircraft platforms, flight control and power systems. The development of the drone industry has shown explosive growth, the market space is growing rapidly, and drones have become the focus of various capital concerns. For example, Snap acquired the drone startup Ctrl Me Robotics, Caterpillar Group made a strategic investment in the U.S. drone service giant Airware, and Intel acquired the German drone software and hardware manufacturer MAVinci.

1.3.2 Chinese Special Robot Development Status

At present, China's special robot market maintains rapid development, with various types of products emerging, and there is a prominent demand for special robots in response to earthquakes, floods and extreme weather, as well as public safety events such as mining accidents, fires and security. In 2016, China's special robot market size reached USD630 million, with a growth rate of 16.7%, slightly higher than the global growth rate of special robots. Among them, the market size of military application robots, extreme operation robots and emergency rescue robots was USD480 million, USD110 million and USD0.4 billion, respectively, with extreme operation robots being the fastest-growing area. As China's enterprises further raise their awareness of safety production, they will gradually use special robots to replace people in high-risk places and complex environments. By 2020, the domestic market demand for special robots is expected to reach $1.24 billion (see Fig. 1.3).

China's special robots have been developed from scratch, enriched in variety and expanded in application areas, laying the foundation for the industrialization of special robots. The Chinese government attaches great importance to the research and development of special robot technology, and supports it through the 863 program, key technology theme projects for robots operating in special service environments and key technologies and equipment for deep sea. At present, some key core technologies in the field of anti-terrorist and detonation-proof and deep-sea exploration have made breakthroughs, such as indoor positioning technology, high-precision positioning navigation and obstacle avoidance technology, and rapid identification technology of dangerous goods in car chassis has been initially applied to anti-terrorist and detonation-proof robots. At the same time, China has overcome a number of core technologies such as titanium alloy manned module spherical shell manufacturing,

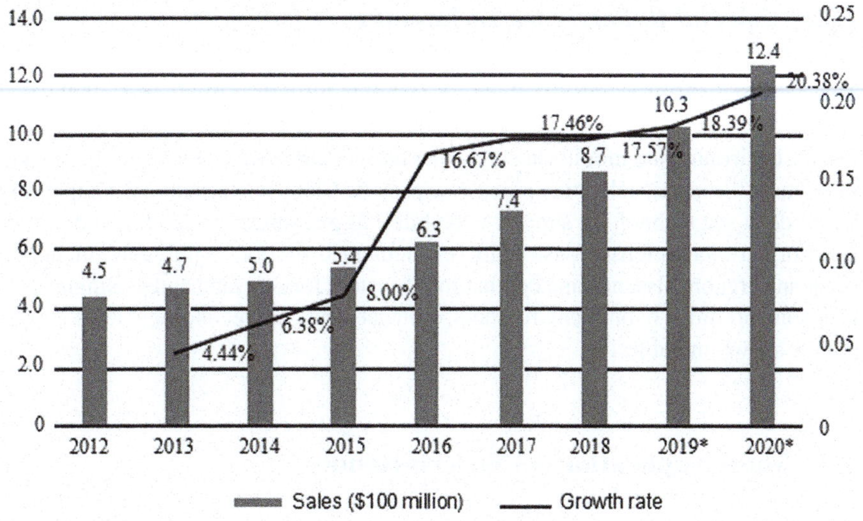

Fig. 1.3 2012–2020 China's special robot sales and growth rate

large depth buoyancy material preparation, deep-sea thrusters, etc., so that China has made significant progress in the localization of deep-sea core equipment.

Over the past 20 years, China has developed a large number of special robots and put them into use, such as auxiliary orthopedic surgery robots and brain surgery robots successfully used in clinical surgery, low-altitude flight robots have been applied in Antarctic scientific research, micro-mine-detecting and minesweeping robots have participated in international peacekeeping minesweeping operations, aerial search and detection robots, ruins search and rescue robots and other earthquake search and rescue robots have been successfully introduced, cell injection Micro-operation robots have been used in animal cloning experiments, the first domestic minimally invasive abdominal surgery robot has been tested on animals and passed the appraisal, anti-terrorist detonation robots have been batch equipped with public security and armed police forces, etc.

Special drones, underwater robots and other development levels of global leadership. At present, in the field of special robots, China has initially formed a series of products such as special drones, underwater robots, search and rescue/explosive robots, and formed advantages in some fields. For example, China Electronics Technology Group Corporation has researched and developed an intelligent cluster system for fixed-wing UAVs, and successfully completed a cluster flight test of 119 fixed-wing UAVs. China's China Railway Times Electric Company developed the world's largest tonnage deep-water trenching plow, filling the gap in the field of China's deep-sea robotics equipment manufacturing; a new generation of ocean-going research vessel "Science" carrying the cable-controlled remotely operated vehicle "Discovery" and autonomous Underwater robot "Exploration" in the northern part of the South China Sea to achieve the first deep-sea rendezvous shooting.

1.3.3 *Special Robots Core Technology*

The development of marine exploration technology with independent intellectual property rights, research and development for resource exploration, fishing and rescue, environmental monitoring and other needs of the series of marine equipment, to promote the industrialization process, to provide China's international competition in the deep sea technology support and ability to guarantee; research and development of national defense construction of unmanned robotics equipment, including unmanned reconnaissance and combat robot system for sea, land and air single environment and multi-habitat environment, enhance The core technology development plan is shown in Table 1.1.

1.4 Main Applications of Special Robots

After decades of development, special robots have been widely used in medical, agricultural, construction, service, military, and disaster relief applications. The following is a brief overview of the many areas in which robots are used.

- Medical robots

A medical robot is a robot that is used in hospitals and clinics for medical treatment or auxiliary medical treatment. It can prepare operation plan alone, determine action program based on actual situation, and then turn the action into movement of operating mechanism. The main research includes planning and simulation of medical surgery, robot-assisted surgery, minimal-injury surgery, and proximity-sensitive surgery. The U.S. has conducted research on proximity sense surgery for battlefield simulation, surgical training, anatomy teaching, etc. France, Britain, Italy, Germany and other countries have jointly conducted research on projects such as image-guided orthopedic surgery programs, pocket robot programs, and electromechanical surgical tools for surgical procedures, and have achieved some success. There are many types of medical robots, and according to their uses, there are clinical medical robots, nursing robots, medical teaching robots, and robots for disabled people.

– Robot for drug delivery

Drug delivery robots can replace nurses to deliver meals, cases and labs, etc. Some of the more famous ones are the Help Mate robot from TRC in the United States.

– Robot for moving patients

Mobile patient robots mainly help nurses move or transport paralyzed, and mobility impaired patients, such as the UK's PAM robot.

– Robotics for clinical care

Table 1.1 Special robots core technology development plan table

Special robot		Status quo	Recent (2020)	Forward (2030)
		Only in a known environment and facing a specific task, can special robots show autonomy in some aspects	The ability of autonomous navigation, guidance and control of robots in complex environment is improved. Robots can get rid of the continuous real-time remote control by human and partially complete some tasks autonomously	Robots are able to cope with environments that require high cognitive abilities (wild natural environments) and operate autonomously without relying on human remote control
Robot ontology technology	Drive technology	Electric, hydraulic, gas and other driving methods are the mainstream, and the driving performance is not high	Driving performance improvement: rapid development of lightweight, miniaturized and integrated technologies	New driving methods (chemical driving, nuclear driving, biological driving) appear and gradually mature
	Institutional configuration	Traditional mechanisms and configurations have low performance in terms of dexterity and efficiency	With the rapid development of bionic mechanism technology, the mechanism performance has been greatly improved	Bionic motion mechanism may exhibit bionic motion performance
Sensing and control technology	Motion control	The motion control technology under conventional conditions has matured, but the motion control performance under complex conditions is still not high	Motion control technology tends to mature, which can support robots to safely complete some complex motions under complex conditions	Robust control and adaptive control technology have been widely used, and the robot can realize most maneuvering operation modes

(continued)

Table 1.1 (continued)

	Perception	Task-oriented perception technologies such as unstructured environment modeling technology and specific target recognition technology are gradually mature	Technologies such as dynamic environment perception and long-term autonomous perception tend to mature, but environmental cognition is still insufficient	The robot's perception ability has been greatly enhanced, the perception accuracy and robustness have been greatly improved, and the robot will have certain situational awareness capabilities
Intelligence and autonomy technology	Navigation planning decision	The navigation and planning technologies for specific missions and environments are mature. In the short term, the decision-making autonomy of robots is still not high, and it needs to rely on operators to make decisions	The handling of uncertain factors in navigation and planning algorithms tends to mature, the real-time performance of the algorithm is greatly improved, and the robot can make decisions for specific tasks	The inherent processing mechanism of system uncertainty in navigation and planning is mature, and real-time navigation and planning enables robots to make autonomous decisions in some complex environments (polar regions, oceans, planets, etc.)
	Learning	The learning theory for specific tasks (such as target recognition, navigation, etc.) tends to mature, which can improve the efficiency of task completion	The theory of autonomous learning is developing rapidly, and robots can realize task-oriented autonomous developmental learning	The development of autonomous learning theory tends to mature, and technologies such as cognitive learning, long-term learning, and machine-robot-machine fully autonomous learning (through observation and interaction) are developing rapidly

Robots for clinical medical use include surgical robots and diagnostic and therapeutic robots. The robot shown in Fig. 1.4 is a medical robot capable of treating patients for stroke, and this robot is capable of interacting physician and patient information via the Internet. With this robot, doctors can visit and treat patients without having to meet them face-to-face.

– Robot for people with disabilities

Robots for people with disabilities, also called rehabilitation robots, can help people with disabilities regain their ability to live independently. The robot shown in Fig. 1.5 is a new type of disabled robot, which is designed by the U.S. military specifically

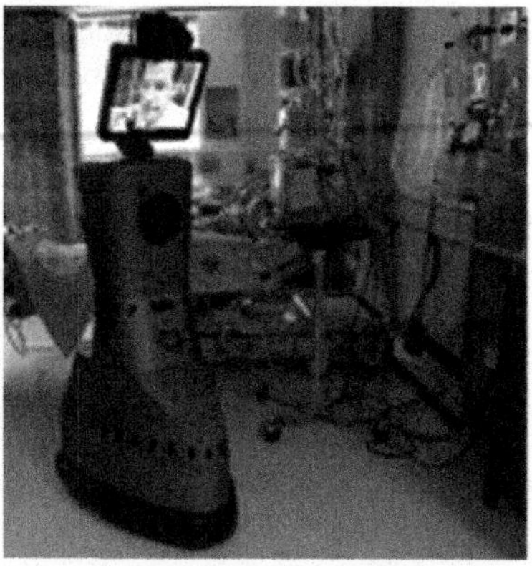

Fig. 1.4 Robotic doctor

Fig. 1.5 Disability assistance robot

for soldiers who have lost mobility due to war injuries. It wraps the injured soldier's lower limbs tightly inside the robot and controls the robot's walking by sensing the soldier's limb movements.

– Nursing Robot

British scientists are developing a nursing robot that can be used to share the heavy and trivial nursing work of caregivers. The newly developed nursing robot will help health care workers to identify patients and distribute the required medication accurately. In the future, the nursing robot can also check the patient's temperature, clean the room, and even help doctors keep abreast of the patient's condition through video transmission.

– Medical teaching robot

Medical teaching robots are ideal teaching aids. U.S. health care workers currently use a teaching robot called "Noel" that can simulate a pregnant woman in labor and even talk and scream. By simulating a real birth, it helps to improve the surgical cooperation and response of obstetricians and gynecologists in the field.

- Agricultural robots

Agricultural robots refers to the general term for robots applied to agricultural production. In recent years, with the development of agricultural mechanization, agricultural robots are playing an increasing role and have been put into application, such as tomato picking robots (see Fig. 1.6), forest tree bulb picking robots (see Fig. 1.7), grafting robots (see Fig. 1.8), root cutting robots (see Fig. 1.9), harvesting robots, spraying robots, etc.

- Construction robots

Construction robot refers to the general term for robots applied in the construction field. With the rapid growth of the global construction industry and the rise in labor costs, construction robots have ushered in a market. 20 variety of construction robots have been developed in Japan. For example, plastering robots for high-rise buildings, prefabricated parts installation robots, interior decoration robots, floor polishing

Fig. 1.6 Tomato picking robot

Fig. 1.7 Forest tree ball fruit picking robot

Fig. 1.8 Grafting robot

Fig. 1.9 Root cutting robot

robots, glass-cleaning robots, etc., have been practically applied. The United States is conducting the development of pipe digging and burial robots, interior wall installation robots and other models, and has carried out basic research on sensors, mobile technology and system automation construction methods. Figure 1.10 is a glass curtain wall cleaning robot, and Fig. 1.11 is a pipe cleaning robot.

Construction robots are widely used abroad because construction robots can work 24 hours per day and maintain a special posture for a long time without knowing fatigue. The house built by the robot is of better quality and can withstand bad weather without problems. A U.S. company has introduced a brick-laying robot called "semi-automatic Mason" (see Fig. 1.12), which can lay 3000 bricks per day, while a domestic worker generally lays only 250–300 bricks per day.

The fully automated commercial construction robot Hadrian X is developed in Australia, which can realize 3D print and lay bricks, and it can lay a staggering 1000

Fig. 1.10 Glass curtain wall cleaning robot

Fig. 1.11 Pipe cleaning robot

Fig. 1.12 "Semi-automatic Mason" (SAM100) bricklaying robot

bricks per hour (see Fig. 1.13). Instead of using traditional cement, Hadrian X uses construction adhesive to bond the bricks, which greatly increases the speed, strength and can reinforce the final thermal effect of the structure.

In April 2018, a team of researchers at MIT developed a new digital construction machine that can use 3D printing technology to "print" building, the robot uses a construction material that is a mixture of foam and concrete, with gaps left between the walls to embed lines and pipes. The bottom unit of this product resembles a detection vehicle fitted with tank tracks, with two robotic arms on top and nozzles at the end of the arms (see Fig. 1.14).

Fig. 1.13 Hadrian X construction robot

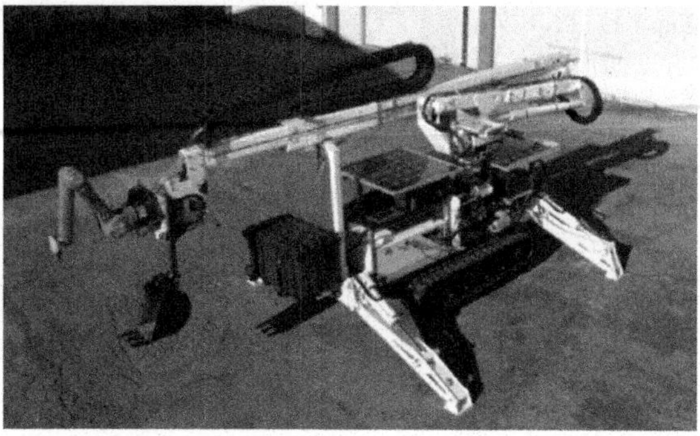

Fig. 1.14 Digital construction machine developed by MIT

Figure 1.15 shows the robot used to bind reinforcing steel on a trial basis on a bridge project in Pennsylvania, USA. The Husqvarna DXR series of remote-controlled demolition robots, shown in Fig. 1.16, are powerful, lightweight and designed with functionality in mind. Workers can operate the Husqvarna DXR demolition robot remotely without having to enter a dangerous demolition site.

“Auto-rovers” robots on construction sites are equipped with high-definition cameras and sensors, which can navigate around the site and can identify and avoid obstacles. The “EffiBOT” robots are developed by the French robotics company Effidence, which can follow workers and carry tools and materials (see Fig. 1.17).

Fig. 1.15 Robot bundling steel bars

Fig. 1.16 Husqvarna DXR demolition robot

Fig. 1.17 EffiBOT developed by French robot company Effidence

- Home service robots

A service robot is a type of robot that operates in an autonomous or semi-autonomous manner and can provide services for human life and health, or a class of robots that can perform maintenance on the operation of equipment. A service robot is mainly a mobile platform that can move, with some arms on it for operation, and is also equipped with some force sensors, vision sensors, ultrasonic distance sensors, etc. It recognizes the surrounding environment, judges its own movement, and completes some kind of work, which are the basic characteristics of service robots.

As shown in Fig. 1.18, Japan invented the humanoid robot nanny “AR”, “AR” equipped with a total of five cameras, through image recognition to identify furniture, it relies on wheels to move, in addition to laundry, cleaning, and clean up, it will also clean up dishes and many other household chores. During the public demonstration, AR demonstrated the process of opening the lid of a washing machine and putting clothes into it, as well as the functions of delivering dishes and cleaning.

The robot shown in Fig. 1.19 is a new generation of robot nanny Care-O-Bot3 developed by Germany, which is covered with a variety of sensors that can identify objects, allowing it to accurately determine the location of objects and will and

Fig. 1.18 Robot Nanny

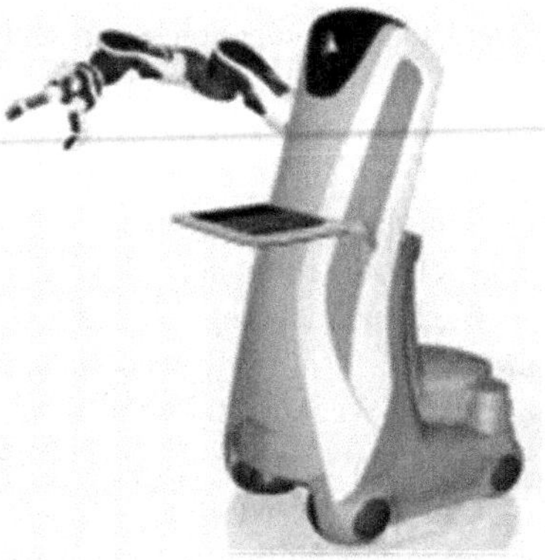

Fig. 1.19 Robot Care-O-Bot3

identify their type; he can not only be controlled by voice or gesture control, it also has a strong self-learning ability.

The robot shown in Fig. 1.20 is a humanoid robot "Aneia", which was developed in Russia using many unique researches of the company "New Era", and is a service robot that can walk on two feet and talk to people. It has the world's most advanced mechanical structure and program security system.

- Entertainment robot

Entertainment robots have entertainment purposes for human viewing and entertainment, which can resemble people, like some kind of animals, like characters in fairy tales or science fiction, etc. The function of the robot, can walk or complete the action, can have language ability, can sing, have some perception ability. Such as robot singers, soccer robots, toy robots, dancing robots, etc.

The basic functions of the entertainment robot mainly use super AI technology, super gorgeous sound and light technology, visual call technology, and custom effect technology. AI technology gives the robot a unique personality and interacts with people through voice, sound and light, action and touch response; super gorgeous

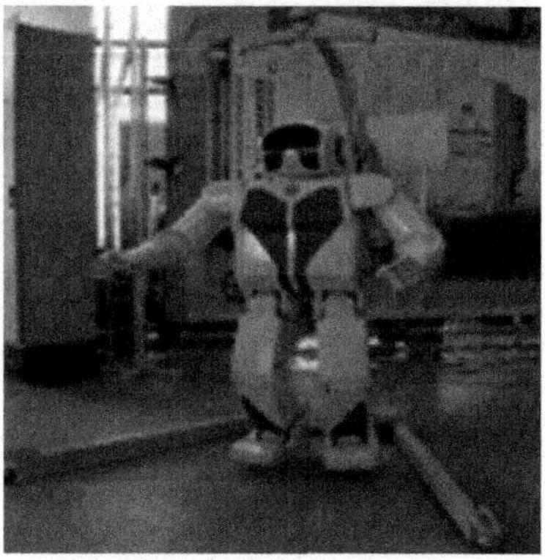

Fig. 1.20 "Aneia" robot

sound and light technology presents super gorgeous sound and light effects through multi-layer LED lights and sound system; visual call technology is to achieve visual call with off-site through the robot's large screen, The visual call technology is realized through the robot's large screen, microphone and speaker, and visual call with foreign places; and the customized effect technology can add different application effects for the robot according to different needs of users.

The break-dancing robot shown in Fig. 1.21 was developed by engineers at RM in the UK. It not only becomes a helper in the classroom to help children learn, but can also control the movement of several joints on its body through a computer programmed to make various movements similar to those of human dancers.

The robot shown in Fig. 1.22 is a "beautiful robot" developed by our scientists, which is not only able to talk with people, but also able to move independently according to the sensors it carries. This "beautiful robot" has a beautiful appearance and can respond quickly to human voice commands.

Sony's new Aibo robot dog (see Fig. 1.23) weighs about 2.2 kg and measures the width and height when standing: 18 cm and 29.3 cm, respectively. The robot itself is

Fig. 1.21 Breakdancing robot

Fig. 1.22 Beauty robot

Fig. 1.23 Sony's new version of the Aibo robot dog

equipped with supercapacitor and 2-axis actuators specially designed by Sony. These actuators enable Aibo's body to move along 22 axes. This allows the new version of Aibo to move more smoothly and naturally than the original Aibo, as reflected in the ear and tail swing, as well as mouth, paw and body movements. The new robot dog is also equipped with a fisheye camera and a camera in the back, both of which work with sensors to detect and analyze sounds and images and help Aibo recognize the face of its owner. Simultaneous localization and mapping technology allows Aibo to adapt to its environment. This combination of sensors and deep learning also helps Aibo analyze praise, interpret smiles and responses to caresses, which creates "a relationship that grows with its owner over time." The SIM card connection provides Aibo with mobile Internet access, which Sony plans to expand to home appliances and devices.

- Military robots

Military robot is a kind of robot with some humanoid function used in military field (reconnaissance, surveillance, detonation, attack, rescue, etc.). In recent years, the United States, Britain, France, Germany and other countries have developed the second generation of military intelligent robots. It is characterized by the use of autonomous control, can complete reconnaissance, combat and logistical support tasks, in the battlefield with the ability to see, smell and touch, can automatically track the terrain and choose the road, and has the function of automatic search, identification and elimination of enemy targets. Such as the U.S. Navplab autonomous navigation vehicle, SSV semi-autonomous ground combat vehicle, France's autonomous rapid-motion reconnaissance vehicle, Germany's MV4 EOD robot, etc. According to its working environment can be divided into ground military robots, underwater military robots, air military robots and space robots, etc.

– Ground military robots

Ground military robots are mainly robotic systems used on the ground, which not only help civilian police to remove bombs and complete security tasks in key locations in peacetime, but also replace soldiers in the battlefield to perform various tasks such as transportation, mine clearance, reconnaissance and attack. There are many

Fig. 1.24 Combat robot

types of ground military robots, mainly combat robots (see Fig. 1.24), explosion-proof robots (see Fig. 1.25), minesweepers, machine security guards (see Fig. 1.26), machine scouts (see Fig. 1.27), etc.

– Underwater military robots

Fig. 1.25 Explosion-proof robot

Fig. 1.26 Machine security

Fig. 1.27 Machine scout

Underwater robots are divided into two categories: manned robots and unmanned robots: among them, manned submersibles are mobile and flexible, easy to deal with complex problems, operator's life may be at risk, and expensive.

An unmanned submersible is what people call an underwater robot. According to the different ways of contact between unmanned submersible and surface support equipment (mother ship or platform), underwater robots can be divided into two categories: one is cable underwater robots, which are customarily called remotely operated vehicles, or ROVs for short; the other is cableless underwater robots, which are customarily called autonomous submersibles, or AUVs for short. Cable robots are all remotely operated, and are divided into towed, (subsea) mobile, and floating (self-propelled) according to their movement. The other type is the cableless robot, which is customarily called autonomous underwater vehicle (AUV). The cableless underwater robots can only be autonomous, only the observation type floating type of movement, but its future is bright. In order to compete for the right to control the sea, all countries are developing underwater robots for various purposes, Fig. 1.28 is the LBV 300 cable-type underwater robot developed by SeaBotix in the United States, and Fig. 1.29 is the CR-01 cable-free autonomous underwater robot developed by China.

Fig. 1.28 LBV 300 cable-mounted underwater robot

Fig. 1.29 CR-01 autonomous underwater robot

– Airborne military robots

Aerial robots, also known as drones, are the most active area in the military robotics family in terms of research activity, technological progress, research and procurement funding, and field experience. Since the first autopilot was introduced, the development of the world's UAVs has basically been moving forward with the U.S. As the main line, and the U.S. ranks first in the world, both in terms of technology level and the types and numbers of UAVs.

UAVs are widely used in reconnaissance, surveillance, early warning, and target attack (see Figs. 1.30, 1.31, and 1.32). With the development of technology, the size of UAVs is getting smaller and smaller, giving rise to micro air vehicles, a product of microelectromechanical system integration. Micro air vehicles are considered to be important reconnaissance and attack weapons on the future battlefield, capable of transmitting real-time images or performing other tasks with sufficiently small size (less than 20 cm), sufficiently large cruise range (e.g., no less than 5 km), and sufficiently long flight time (no less than 15 min).

– Space robot

The space robot is a low-cost, lightweight, remote-controlled robot that can navigate and fly through the atmospheric environment of a planet. For this reason, it must overcome many difficulties, such as it has to be able to move and navigate autonomously in a constantly changing three-dimensional environment; it can hardly stay; it must be able to determine its position and state in space in real time; it has to be able to

Fig. 1.30 Global Hawk UAV

Fig. 1.31 Microstar micro UAV

Fig. 1.32 Machine dragonfly

control its vertical motion; and it has to predict and plan the path for its interplanetary flight. At present, the United States, Russia, Canada and other countries have developed a variety of space robots, such as the Mars robot developed by the United States (shown in Fig. 1.33), the lunar exploration robot (shown in Fig. 1.34), and the International Space Station assembly and maintenance robot (shown in Fig. 1.35). Figure 1.36 shows a Chinese lunar rover conducting desert experiments.

– Disaster rescue robot

In recent years, especially after the "9.11" incident, many countries around the world have started to develop various anti-terrorism and explosion-proof robots, disaster rescue robots and other hazardous work robots from the perspective of national security strategies for disaster protection and rescue. At the same time, due to the potential application background and market for rescue robots, some companies have

Fig. 1.33 U.S. Mars robot

Fig. 1.34 U.S.-developed lunar exploration robot

Fig. 1.35 International space station assembly and repair robots

Fig. 1.36 China's lunar rover conducting desert experiments

also become involved in the research and development of rescue robots. At present, disaster rescue robot technology is developing from theoretical and experimental research to practical application.

The snake-like search-and-rescue robot KOHGA2 developed by the Tokyo Electro-Communication University in Japan is shown in Fig. 1.37, which can enter the narrow space of the disaster site to search for survivors. The use of multiple units in combination increases the freedom of movement of the robot, which can not only prevent it from getting stuck inside the rubble, but also enhance its ability to go over gullies and over obstacles.

(a) Single module (b) Dual module (c) Triple module

Fig. 1.37 Rearrangeable snake-like rescue robot KOHGA2

The rescue robot for nuclear power plant accidents developed jointly by Kobe University and the National Fire and Disaster Research Institute of Japan is shown in Fig. 1.38. It is designed to enter the interior of a contaminated nuclear power facility to transfer fainted survivors to a safe place. This robotic system consists of a group of small mobile robots that operate by first adjusting the body posture of the fainted person for transport through small traction robots, followed by a mobile robot with a stretcher structure to transfer the person to a safe area.

The micro flying robot uFR jointly developed by Chiba University and Seiko Epson Japan is shown in Fig. 1.39. uFR looks like a helicopter and uses an ultra-thin ultrasonic motor with the world's largest power/weight output ratio, weighing only 13 g in total, while uFR can balance in mid-air due to its stable mechanical structure using linear actuators. uFR can be applied in earthquakes and other In natural disasters, it can measure the site as well as the environment in hazardous areas and confined spaces very effectively, and in addition it can effectively prevent secondary disasters.

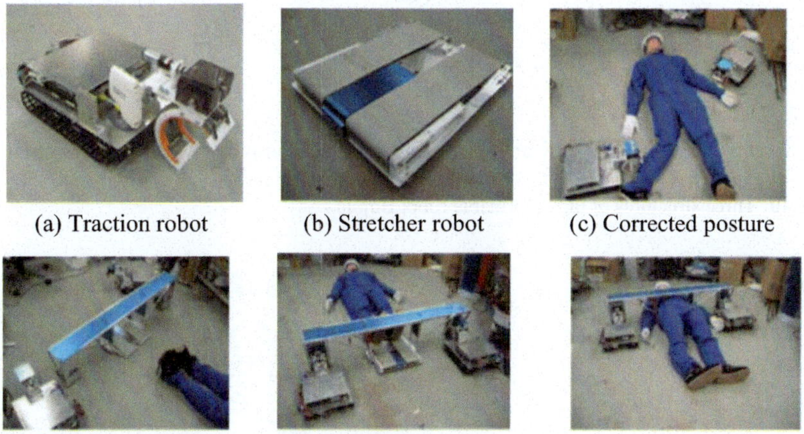

(a) Traction robot (b) Stretcher robot (c) Corrected posture

(d)Robotic connection (e) Fainting person on a stretcher (f) Carrying a fainting person

Fig. 1.38 Rescue robots and their experiments for nuclear power disasters

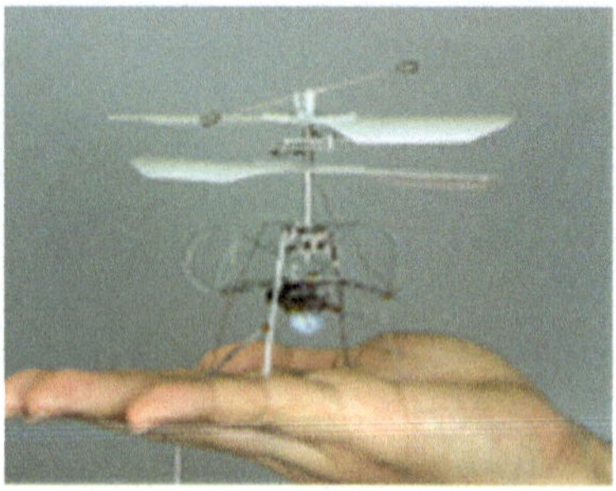

Fig. 1.39 Micro flying robot (uFR)

Bujold, a deformable robot developed by the University of South Florida, is shown in Fig. 1.40. This robot is equipped with medical sensors and cameras, and its bottom is driven by deformable tracks that can be turned into three structures: sitting up and facing forward, sitting up and facing backward, and lying down posture. Bujold has a strong movement and detection ability, which was able to access the disaster site to obtain physiological information about the survivors as well as information about the surrounding environment.

The RoboSimian robot developed by NASA stands upright at 1.64 m and weighs 108 kg (see Fig. 1.41). It has agile and flexible limbs that can move in a quadrupedal manner, and is able to adapt to a variety of complex earthquake debris environments, with good locomotion in the debris environment and strong balance, and is equipped with multiple cameras, capable of acquiring a wealth of information about the outside environment.

(a) Sitting upright facing forward (b) Sitting upright facing backward (c) Lying flat

Fig. 1.40 Three structures of the deformable robot Bujold

Fig. 1.41 RoboSimian robot

Fig. 1.42 CHIMP robot

The CHIMP, a four-limbed robot developed by Carnegie Mellon University (see Fig. 1.42), is a wheel and foot composite mobile robot, the robot's limbs are equipped with a track mechanism, which can adopt the motion of a tracked robot to move on rugged roads, and can also adopt the way the limbs crawl to move, the top of the robot's limbs are equipped with a three-finger manipulator, which can grasp objects, and the limb mechanism and Each joint of the CHIMP robot can be remotely controlled by the operator, and the robot is pre-programmed to perform pre-programmed tasks, with the operator giving high-level commands and the robot performing low-level reflexes and being able to protect itself. The robot has great potential in the field of disaster rescue due to its high ability to adapt to complex environments, its high locomotor capacity and its strong operational capabilities.

The PackBot robot from iRobot (see Fig. 1.43) is a tracked search-and-rescue robot with a forward swinging arm and robotic structure, originally a military security robot, which was deployed to search for survivors in the damaged buildings of the World Trade Centre after 9/11 and recovered a number of survivors. The robot is equipped with a camera in its head to navigate over rough terrain and to change the height of its observation platform, a global positioning system (GPS), an electronic compass and a temperature detector in its chassis, as well as a variety of sensors such as acoustic locators, laser scanners and microwave radar to sense information about the external environment and its own status. The robot has now been developed as a portable mobile control platform based on Android.

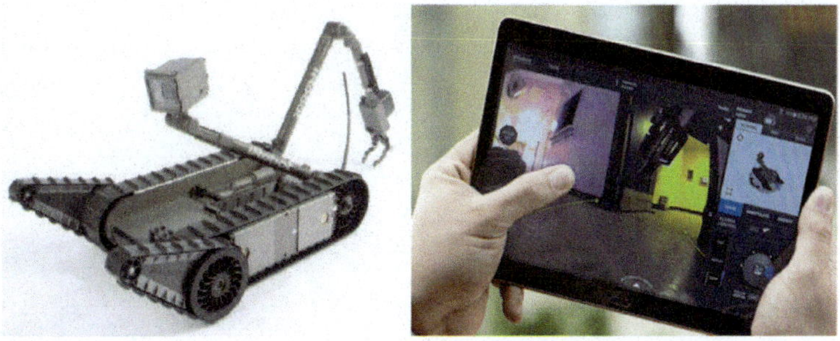

Fig. 1.43 PackBot robot

Fig. 1.44 Honeywell's miniature UAV RQ-16A T-Hawk

The RQ-16A T-Hawk, a vertical takeoff and landing micro UAV developed by Honeywell, is shown in Fig. 1.44. This UAV weighs 8.4 kg, can fly continuously for 40 min, has a maximum speed of 130 km/h, a maximum distance of 3200 m, and a maximum maneuverable range radius of 11 km, and is suitable for backpack deployment and single-person operation. It has been used in the 2011 nuclear power accident in Fukushima, Japan, to help Kyodo Electric Power Company better determine the location of radioactive leaks and how to better handle them.

The portable fire evacuation robot developed by the Gyeongbuk Institute of Science and Technology in Daegu, South Korea, is shown in Fig. 1.37. The evacuation robot is designed to go deep into the fire scene to collect environmental information, find survivors, and guide trapped people to evacuate the fire scene. The robot structure is designed with aluminum composite metal, which is heat and water resistant. The robot has a camera to capture the environment of the fire scene, multiple sensors to detect temperature, carbon monoxide, and oxygen concentration, and a speaker to communicate with the trapped people.

ASGUARD, a hybrid wheel-legged structure developed by the German Artificial Intelligence Research Center, is shown in Fig. 1.45. ASGUARD is a hybrid quadruped outdoor robot inspired by insect movement, and the first generation ASGUARD prototype is directly driven by four legs with one degree of rotational freedom, and ASGUARD is designed with the mission of disaster mitigation as well as urban search and rescue.

Fig. 1.45 Wheel-leg hybrid robot ASGUARD

Fig. 1.46 Evacuation robot

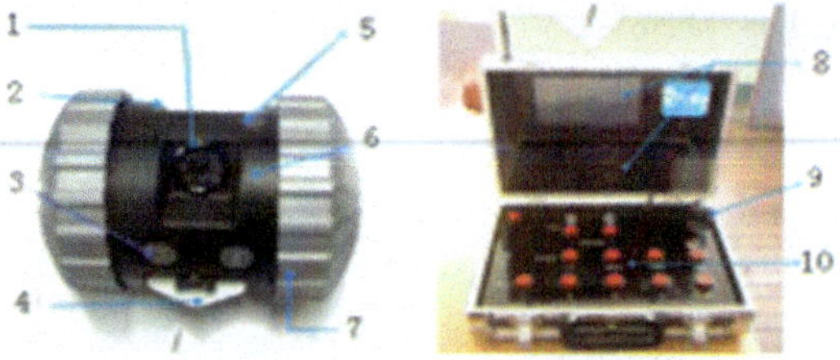

(a) Robot body (b) Control panel

The portable fire evacuation robot developed by the Gyeongbuk Institute of Science and Technology in Daegu, Korea, is shown in Fig. 1.46. The evacuation robot is designed to penetrate the fire scene to collect environmental information, find survivors and guide trapped people out of the fire scene. The structure of the robot is made of aluminium alloy and is heat and water resistant. The robot has a camera to capture the environment of the fire scene, multiple sensors to detect temperature, carbon monoxide and oxygen concentration, and a loudspeaker to communicate with trapped people.

The deformable disaster rescue robot developed by the Shenyang Institute of Automation of the Chinese Academy of Sciences is shown in Fig. 1.47. This robot has nine kinematic configurations and three symmetrical configurations with linear, triangular, and side-by-side forms, and it is able to adapt to the needs of the environment and the task through a variety of forms and gaits, and can be fitted with different equipment such as cameras and life detectors depending on the purpose of use. The deformable disaster rescue robot was first applied in the 2013 earthquake rescue in Lushan County, Ya'an, Sichuan Province, where it was tasked with searching the surface of the rubble and the interior of the rubble to provide the rescue team with the necessary data as well as image support information.

The MicroVGTV robot (see Fig. 1.48), developed by Canadian company Inuktun, is a tracked deformable disaster search and rescue robot with tracks that can be mechanically altered to change the overall structure to suit different conditions

(a) Deformable robot structure diagram (b) On-site disaster relief diagram

Fig. 1.47 Deformable disaster rescue robot and its on-site disaster relief

Fig. 1.48 MicroVGTV robot

and have a high degree of mobility in complex environments. The robot is cable-controlled and fitted with a camera to capture image information of the rubble environment and a micro-microphone and speaker to listen to the sound signals within the rubble, allowing calls to be made to survivors in the rubble.

The Shenyang Institute of Automation, Chinese Academy of Sciences, has developed a seaming robot (see Fig. 1.49), which is a track-driven mobile robot with front and rear swing arms and a front seaming device, mainly used to perform seaming operations on rubble surfaces. The robot is hydraulically driven and has a maximum lifting weight of 1200 kg.

This robot is an automatic propulsion system with a flexible body. The robot consists of an active section and a passive section, with the active section having three degrees of freedom for robot propulsion and robot attitude control, and the passive section can be extended and equipped with internal communication lines and power lines for communication. The passive section can be extended. The robot is equipped with LED lights, a camera, microphone and loudspeaker for lighting and audio communication inside the ruins and to obtain information about the environment inside the ruins, as well as temperature and carbon dioxide sensors to detect the state of the air inside the ruins. The unique mechanism design allows the robot to be highly mobile in a crevice environment and has strong application prospects.

Shanghai University has developed an active-intervening debris crevice search and rescue robot (see Fig. 1.50), which is an automatic propulsion system with a flexible body. The robot consists of two parts: an active segment with three degrees

Fig. 1.49 Seam-raising robot

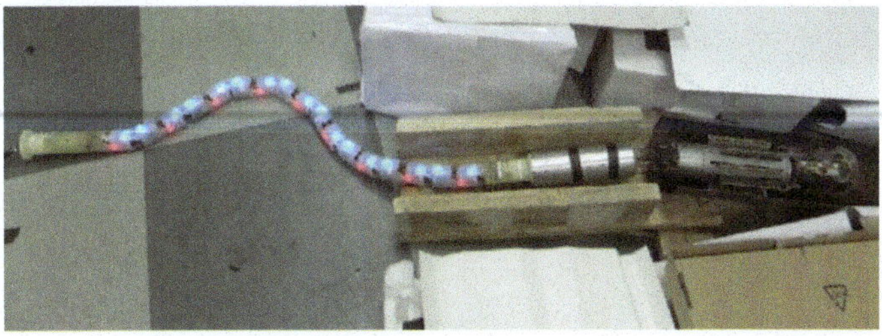

Fig. 1.50 Active intervention crevice search and rescue robot

Fig. 1.51 Rotary-wing flying robot

of freedom for robot propulsion and robot attitude control, and a passive segment that can be extended and equipped with internal communication lines and power lines to play a communication role the passive section can be extended. The robot is equipped with LED lights, a camera, microphone and loudspeaker for lighting and audio communication inside the ruins and to obtain information about the environment inside the ruins, as well as temperature and carbon dioxide sensors to detect the state of the air inside the ruins. The unique mechanism design allows the robot to be highly mobile in a debris crevice environment and has strong application prospects.

The rotary-wing flying robot developed by the Shenyang Institute of Automation of the Chinese Academy of Sciences is shown in Fig. 1.51. The rotary-wing flying robot is small, lightweight, low-altitude, and slow, and is able to overcome factors that are difficult for large aircraft to cope with, such as climate, airflow, and terrain. In the rescue process, the rotary-wing flying robot can obtain the road conditions of the disaster site and the distribution of buildings after the disaster from the air, it can search and investigate by hovering, and transmit high-resolution pictures and images to the operator in real time, providing a basis for decision making for targeted deployment and rescue by rescuers, thus greatly improving the efficiency of disaster rescue.

Chapter 2
Drive System and Mechanism of Special Robot

2.1 Basic Components of Robot

The robot mainly consists of drive system, mechanism, sensing system, human–computer interaction system, and control system, as shown in Fig. 2.1.

- Drive system

The drive system is a device that provides power to the mechanical structure system. The main drive methods of the drive system are: electrical drive, hydraulic drive, pneumatic drive and new drive.

Electrical drive is currently the most used a drive method, which is characterized by no environmental pollution, high accuracy of movement, easy access to power, fast response, high driving force, convenient signal detection, transmission, processing, and can be used in a variety of flexible control methods, the drive motor is generally used stepper motor, DC servo motor, AC servo motor, there are also direct drive motor.

Hydraulic drive can obtain a large gripping capacity, smooth transmission, compact structure, good explosion-proof, action is also more sensitive, but high requirements for sealing, not suitable for high and low temperature site work.

Pneumatic driven robots are simple in structure, fast in action, easy to source air, and low in price, but poor in stability of working speed and low in gripping force due to compressible air.

With the development of applied material science, some new materials are beginning to be used in robot actuation, such as shape memory alloy actuation, piezoelectric effect actuation, artificial muscle and optical actuation.

- Mechanism

The mechanism of the robot consists of transmission mechanism and mechanical components.

T. Guo et al., *Special Robot Technology*, Advanced and Intelligent Manufacturing in China, https://doi.org/10.1007/978-981-99-0589-8_2

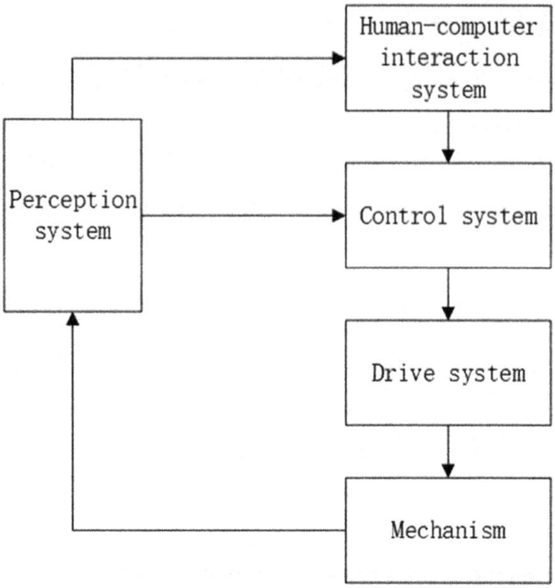

Fig. 2.1 Robot system composition

Transmission mechanisms are used to transfer the motion of the drive to the joints and action parts. Commonly used transmission mechanisms for robots include ball screws, gears, belts and chains, harmonic reducers, etc.

The mechanical components consist of three major parts: the body, the arm, and the end operator. Each large member has several degrees of freedom, constituting a multi-degree of freedom mechanical system. If the base has a movement mechanism, it constitutes a mobile robot; if the base does not have a movement and waist rotation mechanism, it constitutes a single robot arm. The arm generally consists of an upper arm, a lower arm and a wrist. The end-effector is an important component mounted directly on the wrist, which can be a two-finger or multi-finger hand claw, or a work tool.

- Perception System

The sensing system consists of an internal sensor module and an external sensor module to acquire useful information from the internal and external environment. Internal sensors are used to detect the robot's own state (internal information), such as the movement state of joints. External sensors are used to sense the external world and detect the state of the operating object and the operating environment (external information), such as vision, hearing, touch, etc. The use of intelligent sensors improves the mobility, adaptability and intelligence of the robot. The human sensory system is extremely ingenious for perceiving information about the external world, however, for some special information, sensors are more effective than the human sensory system.

- Human–Robot Interaction System

The human–robot interaction system is a device for human–robot contact and participation in robot control. Examples include command consoles, information display panels, and danger signal alarms.

- Control System

The task of the control system is to command the robot's actuators to perform the required movements and functions based on the robot's operating instructions and the signals returned from the sensors.

2.2 Common Drives

The three basic types of drive mode commonly used in robots are hydraulic drive, pneumatic drive and electrical drive. Comparison of drive performance of three drive systems, as shown in Table 2.1.

In the early days of industrial robots, most of them used hydraulic and pneumatic drive methods because most of their movements used crank mechanisms and linkage mechanisms, etc. However, with the demand for high operating speeds and the increasing complexity of the role, the proportion of electrically driven robots is now increasing. However, hydraulic and pneumatic drives are still used satisfactorily

Table 2.1 Comparison of drive performance of three drive systems

Mode	Hydraulic	Electric	Air pressure
Advantages	• Suitable for large robots and large loads • The system has good rigidity, high precision and fast response speed • No reduction gear required • Easy to work in a wide speed range • Can stop in one position without damage	• Applicable to all sizes of robots • Good control performance, suitable for high-precision robots • Compared with hydraulic system, it has higher flexibility • Using the reduction gear to reduce the inertia on the motor shaft • No leakage, reliable and easy maintenance	• High reliability of components • No leakage, no spark • Low price and simple system • Compared with the hydraulic system, the pressure is lower • Flexible system
Shortcoming	• Leakage, not suitable for use in clean occasions • Need pump, liquid storage box, motor, etc. • Expensive, noisy, need maintenance	• Low stiffness • Reduction gear is required, which increases cost, quality, etc. • When the power is not supplied, the motor needs a braking device	• The system is noisy and needs air compressors and filters • It is difficult to control the linear position • Easy to deform under load and low stiffness

in applications where a large amount of force is required, or where motion accuracy is not high and explosion-proof is required.

2.2.1 *Hydraulic Drive*

Hydraulic drives use high pressure oil as the working medium. The drive can be closed-loop or open-loop and can be linear or rotary. Open-loop control enables precise point-to-point control, but cannot stop in between, as it moves from one position and hits a block before stopping.

- Linear hydraulic cylinder

Linear hydraulic cylinders controlled by solenoid valves are the simplest and cheapest open-loop hydraulic drives. In the operation of a linear hydraulic cylinder, the flow is regulated by a controlled throttle, which allows deceleration before reaching the end of the movement, allowing the stopping process to be controlled. There are also many devices that are controlled with manual valves, in which case the operator becomes part of a closed-loop system and thus is no longer an open-loop system. Truck cranes and forklifts are of this type.

Large diameter hydraulic cylinders are expensive, but can output a large force in a small space. The working pressure is usually up to 14 MPa, so it can output 1400 N force per 1 cm^2 area.

Figure 2.2 shows a simplified schematic of a hydraulic cylinder controlled with a servo valve. Whether it is a linear hydraulic cylinder or a rotary hydraulic motor, their operating principle is based on the action of high pressure on the piston or on the vanes. The hydraulic fluid is sent to one end of the cylinder via a control valve, see Fig. 2.2. In an open-loop system, the valve is controlled by a solenoid; in a closed-loop system, it is controlled by an electro-hydraulic servo valve or a manual valve. The most popular Unimation robots have used hydraulic drives for many years.

- Rotating hydraulic motor

Figure 2.3 shows a rotating hydraulic motor. Its housing is made of aluminum alloy, the rotor is made of steel, and the seal and dust ring prevent the oil from leaking and protect the bearing respectively. Under the control of electro-hydraulic valve, hydraulic oil flows in through the inlet hole and acts on the blades fixed on the rotor to make the rotor rotate. The fixed vanes prevent the hydraulic oil from short-circuiting. Position information is given by a potentiometer and a solver driven by a pair of backlash eliminating gears. The potentiometer gives a rough value and the exact position is determined by the solver. In this way, the high precision small range of the solver is compensated by a low precision large range potentiometer. Of course, the overall accuracy does not exceed the accuracy of the gear train driving the potentiometer and the solver.

- **Advantages and disadvantages of hydraulic drive**

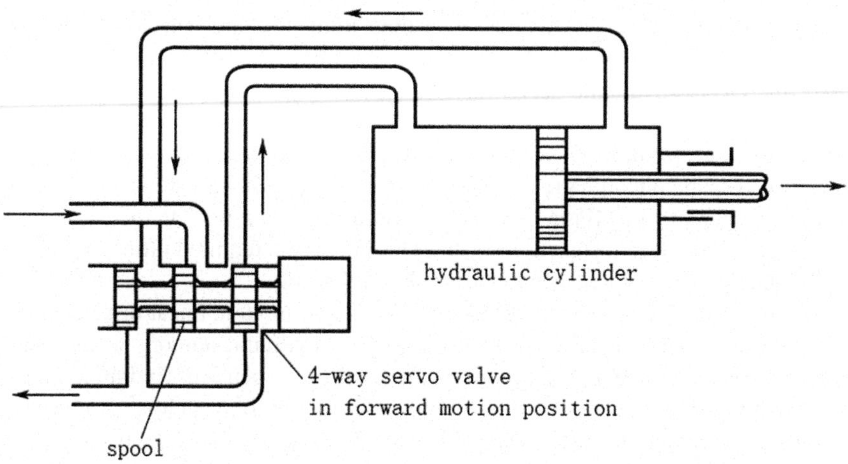

Fig. 2.2 Schematic diagram of a hydraulic cylinder controlled by a servo valve

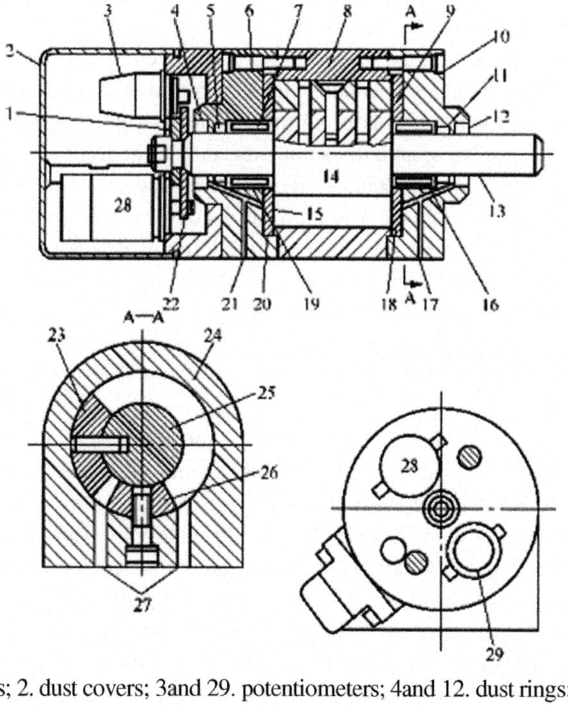

1 and 22. gears; 2. dust covers; 3and 29. potentiometers; 4and 12. dust rings; 5, 11. seals; 6and 10. end caps; 7and 13. output shaft;8 and 24. casing; 9 and 15. steel disc;14 and 25. rotor;16 and 19. needle bearing;17 and 21. drain hole; 18and 20. O-ring; 23. rotating blade; 26. fixed blade; 27. inlet and outlet holes; and 28. solver

Fig. 2.3 Rotating hydraulic motor

Electro-hydraulic servo valves for controlling fluid flow are quite expensive and require filtered, high cleanliness oil to prevent the servo valves from clogging. When used, the electro-hydraulic servo valve is driven by a low-power electrical servo unit (torque motor). The torque motor is relatively cheap, but this cheapness does not compensate for the expensive servo valve itself, nor does it compensate for the defect of system contamination. Because of the high pressure, there is always a risk of oil leakage, and 14 MPa will quickly cover a large area with an oil film, so this is a problem that must be taken seriously. This makes the required fittings expensive and requires good maintenance to ensure their reliability.

Because hydraulic cylinders provide precise linear motion, linear drive elements are used on robots whenever possible. However, hydraulic motors are also well constructed and designed, even though they are more expensive. Hydraulic motors of the same power are smaller in size than electric motors, which is an advantage when hydraulic motors must be installed on the joints of articulated robots. However, for this purpose it is necessary to send hydraulic fluid to the rotary joint. New designs of motors have now become compact in size and reduced in mass because of the use of new magnetic materials. Although more expensive, the motors are still more reliable and require less maintenance.

The fundamental advantage of hydraulic drive over electric drive is its intrinsic safety. In an environment such as painting, stringent demands are placed on safety.

Because there is the possibility of arcing and detonation, the requirement to carry the voltage in the explosive area does not exceed 9 V, the hydraulic system does not have arcing problems, and when used in explosive gases, without exception, always choose hydraulic drive. If an electric motor is used, it must be sealed, but the current cost and quality of the motor for the need for such power is not allowed.

2.2.2 Pneumatic Drive

There are a number of robot manufacturers who build very flexible robots with pneumatic systems. In principle, they are very much like hydraulic drives, but the details are very different. Its working medium is high pressure air. Of all the drive methods, pneumatic drive is the simplest and is widely used in industry. Pneumatic actuators are both linear cylinders and rotary pneumatic motors.

Most of the air pressure drives are completed with inter-block movements. Achieving precise control is difficult due to the compressibility of air. Even if high pressure air is applied to both ends of the piston, the inertia of the piston and load will continue to move the piston until it hits the mechanical block or until the air pressure finally balances with the inertial force.

Achieving high precision with pneumatic servo is difficult, but pneumatic drive is the lightest in mass and the lowest in cost among all robots when accuracy can be met. Precise positioning in point operation can be achieved with mechanical stopper, and 0.12 mm accuracy is easily achieved. A buffer added to the pneumatic cylinder and the stopper can slow down the cylinder at the end of the movement to prevent touching the equipment. The simplicity of operation is one of the main advantages of the pneumatic system. Because it is simple, straightforward and easy to program, it can perform the task of a large number of point handling operations. Point handling is to grab an item from one location and move it to another designated location to put it down.

A new type of pneumatic motor—a vane motor directly controlled by a microprocessor—is able to carry a load of 215.6 N while obtaining a high positioning accuracy (1 mm). The main advantage of this technology is its low cost. Compared to hydraulically and electrically driven robots, pneumatic drives are very competitive if they can achieve high accuracy and reliability.

The greatest advantage of pneumatic drives is their modularity. Since the working medium is air, it is easy to connect many compressed air pipes to the individual drives and to build up an arbitrarily complex system using standard components.

The power source for the pneumatic system is provided by a high quality air compressor. This air source can be shared by all pneumatic modules through a common multi-connector. A solenoid valve mounted on the multi-connector controls the air flow to each pneumatic component. In the simplest systems, the solenoid valves are controlled by step switches or part sensing switches. Several actuators can be assembled to provide 3–6 individual motion.

Pneumatic robots can also be taught like other robots, and point operations can be controlled with the teach box.

2.2.3 Electrical Drive

Electrical drives use the force or torque generated by an electric motor to drive the robot directly or through a reduction mechanism to achieve the position, speed and acceleration required by the robot. Electrical drives do not require energy conversion, are easy to use, have many advantages, such as no environmental pollution, flexible control, high accuracy, low cost and high drive efficiency, and are most widely used. Electrical drives can be divided into stepper motor drive, linear motor drive and servo motor drive.

The speed and displacement size of the stepper motor drive can be controlled by the number of pulses issued by the electrical control system. As the displacement of the stepper motor is strictly proportional to the number of pulses, the stepper motor drive can achieve high repeat positioning accuracy, but the speed of the stepper motor can not be too high, the control system is also more complex.

Linear motor drive structure is simple, low cost, its action speed and travel depends mainly on the length of its stator and rotor, reverse braking, positioning accuracy is low, must be equipped with additional buffer and positioning mechanism.

Servo motor drive according to its use of different power nature, can be divided into DC servo motor drive and AC servo motor drive two types. DC servo motor has the advantages of good speed regulation characteristics, large starting torque, fast response, etc. AC servo motor is simple in structure, reliable in operation and easy to maintain. With the rapid development of microelectronics, AC drive technology, which was mainly used for constant speed operation in the past, gradually replaced the high-performance DC drive in the 1990s, making the maximum speed of the servo actuator of the robot, capacity, operating environment and maintenance and repair conditions have been substantially improved, so as to achieve servo motor performance with lightweight and small in size, easy to install, high efficiency, high control performance, which can meet the requirements of the robot. The AC servo motor used in the robot is basically the same as the DC servo motor, but the only difference is the rectifier part. DC brushed motors cannot be used directly in environments requiring explosion protection, and the costs are higher than those of the two drive systems. However, because of the advantages of these drive systems, they are widely used in robotics.

2.2.4 New Drive

With the development of robotics, new types of actuators have emerged, such as piezoelectric actuators, electrostatic actuators, shape memory alloy actuators, ultrasonic actuators, artificial muscles, optical actuators, etc.

- Piezoelectric actuators

A piezoelectric material is a material that has an electrical charge on its surface proportional to the external force when it is subjected to a force, also known as a piezoelectric ceramic. In turn, when an electric field is applied to a piezoelectric material, the piezoelectric material generates a strain and outputs a force. This property can be used to make piezoelectric actuators, which can be driven with sub-micron accuracy.

- Electrostatic actuators

Electrostatic actuators use the suction and repulsion forces between charges to actuate electrodes in a sequential manner to produce translational or rotational motion. Because electrostatic forces are surface forces, they are proportional to the quadratic side of the component size and can generate large amounts of energy at small changes in size.

- Shape-memory alloy drives

Shape-memory alloys are special alloys that once they have been made to remember any shape, even if they are deformed, when heated to a suitable temperature, they can be restored to their pre-deformed shape. There are dozens of known shape memory alloys such as Au–Cd, In–Tl, Ni–Ti, Cu–Al–Ni, Cu–Zn–Al, etc.

- Ultrasonic actuator

The so-called ultrasonic actuator is a kind of actuator that uses ultrasonic vibration as the driving force, i.e. it consists of a vibrating part and a moving part, and is driven by the friction between the vibrating part and the moving part.

The ultrasonic actuator has no iron cores or coils, is simple in structure, small in size, light in weight, fast in response, and has a large torque, and can operate at low speeds without the need for a reduction device.

- Artificial muscles

With the development of robotics, the drive has evolved from the traditional mechanical movement mechanism of motor-reducer to a biological movement mechanism of skeleton → tendon → muscle. The human arm is capable of performing a wide range of supple tasks, and the actuators developed to achieve part of the skeleton → muscle function are called artificial muscle actuators. Many different types of artificial muscles have been developed to better simulate the motor functions of living organisms or to be used in robots, such as polymer gels using mechanical chemicals and artificial muscles made of shape memory alloys.

- Light-driven

A strong dielectric (tightly asymmetrical piezoelectric crystals) irradiated by light produces a light-induced voltage of several kilovolts per centimetre. This phenomenon is the result of the piezoelectric and photostrictive effects. This is caused by the presence of impurities within the dielectric, resulting in a tight asymmetry of the crystals, which causes a charge shift during the light excitation process.

2.3 Common Transmission Mechanism

The drive mechanism is used to transfer the motion of the drive to the joints and action parts. The common drive mechanisms used in robots are screw drives, gear drives, helical drives, belt and chain drives, linkage and cam drives, etc.

2.3.1 Linear Drive Mechanism

- Screw drive

There are sliding type, ball type and hydrostatic type of screw drive. The screw for robot drive has the features of compact structure, small clearance and high transmission efficiency.

– Ball screw

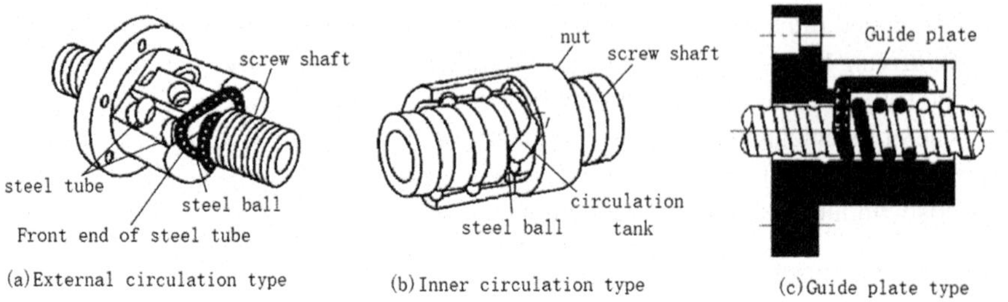

Fig. 2.4 Structure of ball screw

The ball screw is equipped with many steel balls between the screw and the nut, and the balls are continuously circulated when the screw or the nut moves, and the motion is transmitted. Therefore, even if the lead angle of the screw is small, the transmission efficiency of more than 90% can be obtained.

Ball screw can convert linear motion into rotary motion or rotary motion into linear motion. According to the ball screw circulation method, the ball screw is divided into the external circulation method of the ball tube, the internal circulation method of the ball circulation by the S-shaped groove inside the nut and the guide plate method of the ball circulation by the upper guide plate of the nut, as shown in Fig. 2.4.

The linear feed speed obtained from the screw speed and lead is

$$v = 60 \cdot l \cdot n \tag{2.1}$$

where v is the linear motion speed, m/s; l is the lead of the screw, m; n is the speed of the screw, r/min.

The driving torque is given by Eqs. (2.2) and (2.3).

$$T_{\mathrm{a}} = \frac{F_{\mathrm{a}} \cdot l}{2\pi \cdot \eta_1} \tag{2.2}$$

$$T_{\mathrm{b}} = \frac{F_{\mathrm{a}} \cdot l \cdot \eta_2}{2\pi} \tag{2.3}$$

where T_{a} for the rotary motion to linear motion (positive motion) when the driving torque, N m; η_1 for positive motion when the transmission efficiency (0.9–0.95); T_{b} for linear motion to rotary motion (reverse motion) when the driving torque, N m; η_2 for reverse motion when the transmission efficiency (0.9–0.95); F_{a} for the axial load, N; l for the lead of the screw, m.

- Planetary wheel type screw

The planetary wheel type screw is mostly used for high speed feeding of precision machine tools, and from the viewpoint of high speed and high reliability, it can also be used in the transmission of large robots, the principle of which is shown in Fig. 2.5. The nut and the screw shaft have a planetary wheel that engages with the screw shaft,

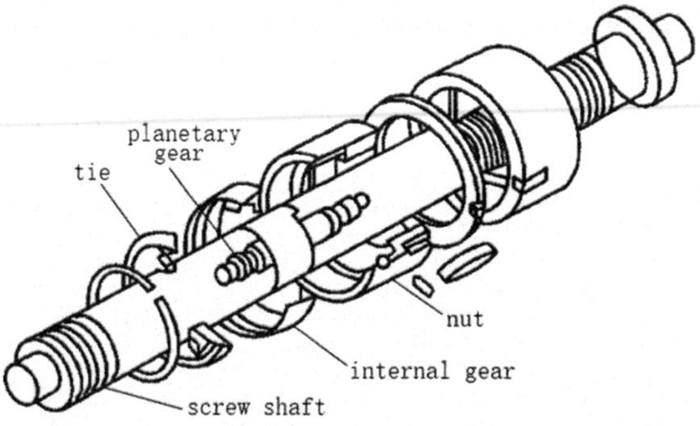

Fig. 2.5 Planetary wheel type screw

and the tether with seven to eight sets of planetary wheels can freely rotate inside the nut. When the nut is fixed and the screw shaft is driven, the planetary wheels rotate on their own and in relation to the internal gear, and the screw shaft moves in the axial direction. The planetary wheel type screw has the advantages of high load capacity, high rigidity and high rotary accuracy, and high accuracy of screw positioning due to small pitch.

- **Belt drive and chain drive**

Belt and chain drives are used to transmit rotary motion between parallel axes or to convert rotary motion into linear motion. Belt and chain drives in robots transmit rotary motion via pulleys or sprockets, respectively, and are sometimes used to drive pinions between parallel shafts.

– Toothed belt drive

Toothed belt has trapezoidal teeth on the driving surface that engage with the pulley, as shown in Fig. 2.6. Toothed belt transmission without sliding, the initial tension is small, and the bearing of the driven shaft is not easily overloaded. Because there is no sliding, it is used for positioning in addition to power transmission. The toothed belt uses neoprene as the base material and adds materials with high expansion and contraction rigidity such as glass fiber in the middle, and the tooth surface is covered with nylon cloth with good wear resistance. Toothed belts used to transmit light loads are made of polyurethane. The pitch of the teeth is expressed in terms of the circular pitch *p of the* envelope pulley, expressed in terms of the modulus method and the inch method. Toothed belts with various pitches are available in different widths and lengths. Let the speed of the active and driven pulleys be n_a and n_b respectively, the number of teeth are z_a and z_b respectively, and the transmission ratio of the toothed belt transmission is

$$i = \frac{n_b}{n_a} = \frac{z_a}{z_b} \tag{2.4}$$

Let the circular pitch be p and the average speed of the toothed belt be

Fig. 2.6 Toothed belt shape

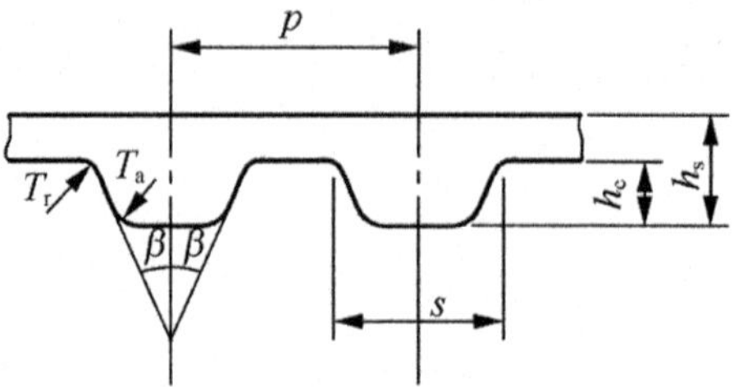

$$v = z_a \cdot p \cdot n_a = z_b \cdot p \cdot n_b \tag{2.5}$$

The transmission power of the toothed belt is

$$P = F \cdot v \tag{2.6}$$

where P *is the* transmission power, W; F is the tight edge tension, N; v is the belt speed, m/s.

Toothed belt drives are low inertia drives, suitable for use between motors and high speed ratio reducers. The belt is mounted on top of the sliding seat to perform the same function as the rack and pinion mechanism. Because it has low inertia and certain stiffness, it is suitable for light sliding seat with high speed movement.

- Roller chain drive

Roller chain drive belongs to a relatively perfect transmission mechanism, which is widely used because of low noise and high efficiency. However, the collision between the roller and the sprocket during the high speed movement produces large noise and vibration, and the satisfactory effect can be obtained only at low speed, that is, it is suitable for the joint transmission with low inertia load. The number of teeth of the chain wheel is small, friction will increase, to get smooth movement, the number of teeth of the chain wheel should be greater than 17, and try to use an odd number of teeth.

2.3.2 Rotational Motion Mechanism

- Types of gears

Gears transmit torque by direct contact of teeth evenly distributed on the wheel edge. Normally, the angular velocity ratio of the gear and the relative position of the shaft are fixed. Therefore, the gear teeth are distributed at equal intervals around the circumference with the contact column surface as the knuckle surface. There are several types of gears depending on the relative position of the shafts and the direction of motion, the main types of which are shown in Fig. 2.7.

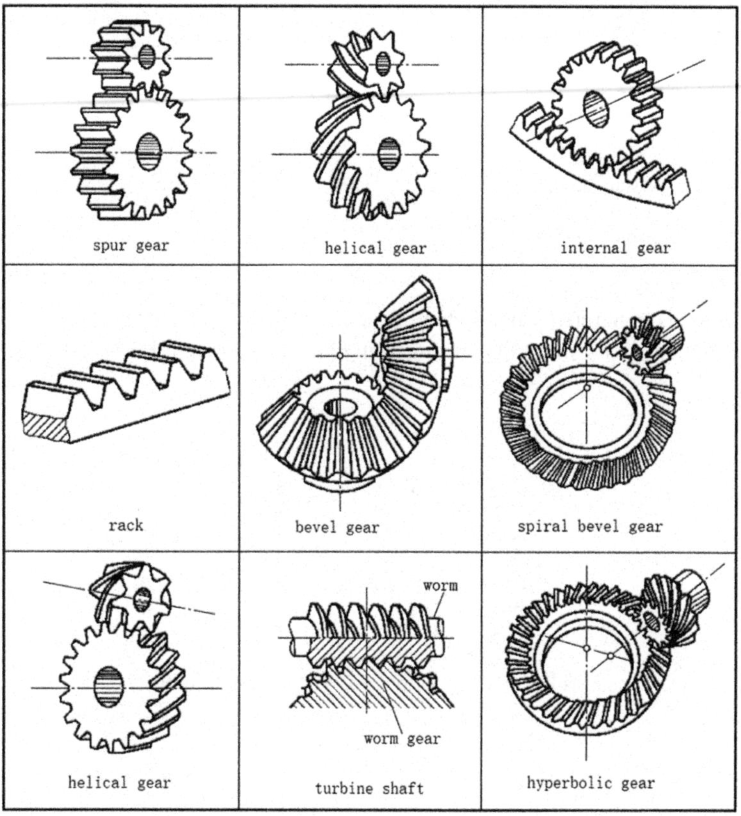

Fig. 2.7 Types of gears

- Structure and characteristics of various gears

– Spur gear

Spur gears are one of the most commonly used gears. Usually, there is a gap between the tooth surfaces at the mesh of two gear teeth, called the tooth gap (see Fig. 2.8). To compensate for gear manufacturing errors and the impact of thermal expansion caused by temperature rise in gear movement, gear transmission is required to have appropriate tooth clearance, but the gear tooth clearance of frequent forward and reverse rotation should be limited to the minimum range. Tooth gap can be adjusted by reducing the tooth thickness or pull up the center distance. The gear mesh without tooth gap is called gapless mesh.

– Helical gear

As shown in Fig. 2.9, the helical gear has twisted tooth bands. It has the advantages of high strength, large overlap coefficient and low noise compared with spur gears.

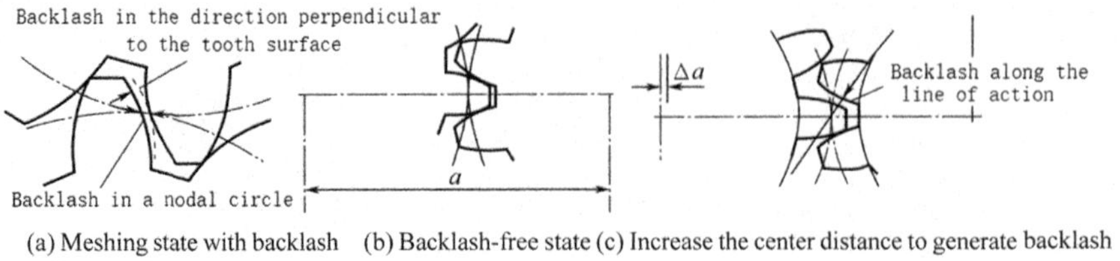

Fig. 2.8 Tooth gap of spur gear

Axial force is generated when helical gears are driven, so thrust bearings should be used or helical gears should be arranged in pairs, as shown in Fig. 2.10.

– Bevel gear

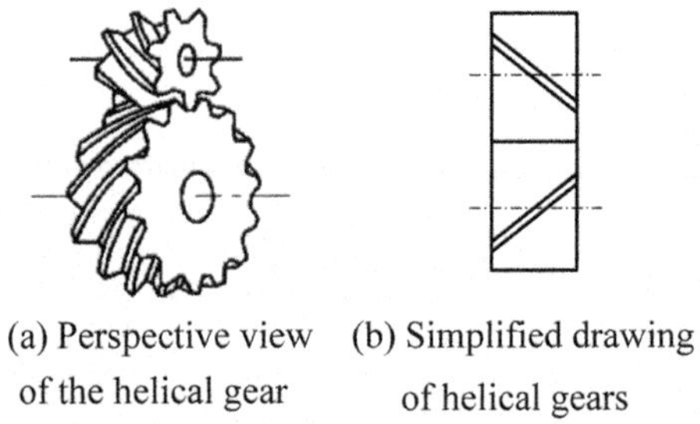

Fig. 2.9 Helical gear

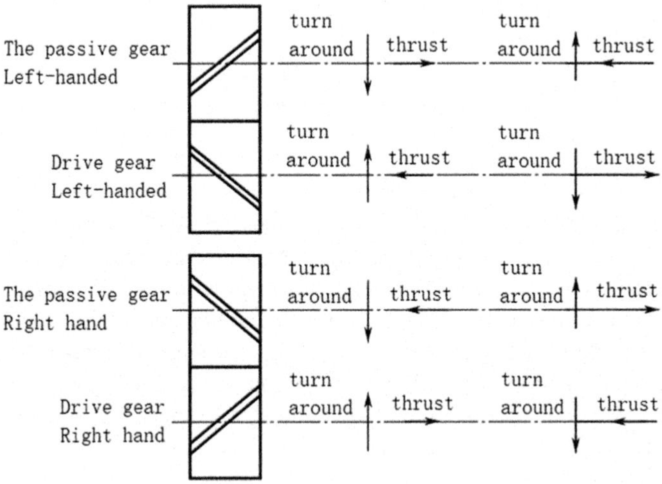

Fig. 2.10 Rotation direction and thrust of helical gear

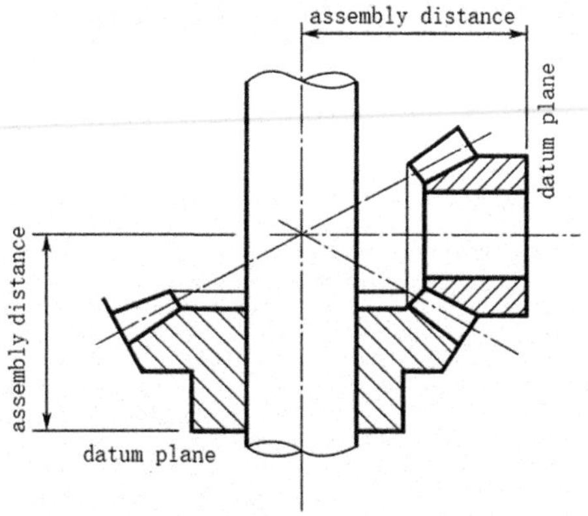

Fig. 2.11 Meshing state of bevel gear

Bevel gears are used to transmit the motion between intersecting axes, and the two conical surfaces with the intersection point of the two axes as the vertex are the meshing surfaces, as shown in Fig. 2.11. The straight bevel gear is called straight bevel gear if the tooth direction is the same as the straight bus of the knuckle bevel, and the curved bevel gear if the tooth direction is in the tangent plane of the knuckle bevel. The straight bevel gear is used for the occasion that the circumferential speed of the knuckle is lower than 5 m/s, and the curved bevel gear is used for the occasion that the circumferential speed of the knuckle is higher than 5 m/s or the speed is higher than 1000 r/min, and it is also used for the occasion that requires smooth rotation at low speed.

– Worm gear

The worm gear transmission consists of a worm gear and a worm wheel that engages with the worm. The worm gear is capable of transmitting motion between vertical shafts with a large reduction ratio. Drum worm gears are used in applications with large loads and large overlap coefficients. Compared with other gear transmission, worm gear transmission has the advantages of low noise, light rotation and large transmission ratio, but the disadvantage is that its tooth gap is larger than spur gears and helical gears, and the friction between the teeth is large, so the transmission efficiency is low.

Based on the characteristics of the various gears mentioned above, the gear transmission can be divided into the types shown in Fig. 2.12. The corresponding type can be selected according to the relative position and steering between the active shaft and the driven shaft.

- Speed ratio of the gearing mechanism

– Optimal speed ratio

A prime mover with limited output torque is required to accelerate the load in a short time, and the speed ratio of its gearing mechanism is required to be optimal.

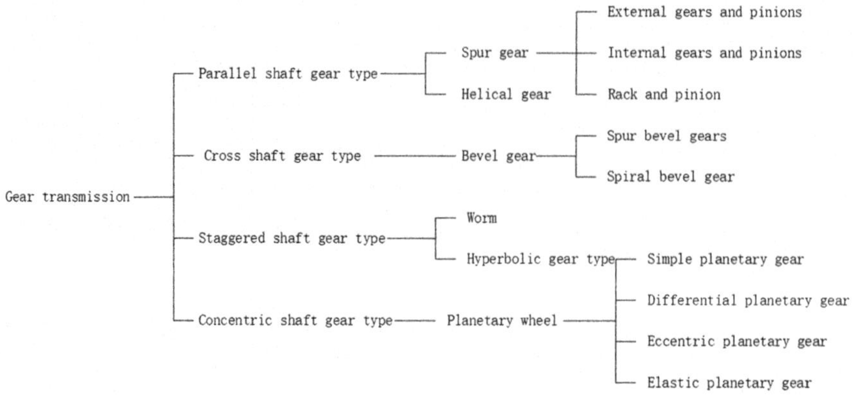

Fig. 2.12 Types of gearing

The prime mover drives the inertia load, and let its moments of inertia be J_N and J_L respectively, then the optimal speed ratio is

$$U_a = \sqrt{\frac{J_L}{J_N}} \tag{2.7}$$

– The number of transmission stages and the distribution of speed ratio

Multi-stage transmission should be used when large speed ratios are required. The distribution of transmission stages and speed ratios is determined according to the type of gear, structure and speed ratio relationship. The usual relationship between the number of transmission stages and speed ratio is shown in Fig. 2.13.

2.3.3 *Speed Reduction Drive Mechanism*

The common gearing mechanisms used in robots are planetary gearing mechanism and harmonic gearing mechanism. The motor is a high speed, small torque drive, while the robot usually requires low speed and high torque, so the planetary gear mechanism and harmonic drive mechanism reducer are commonly used to complete the speed and torque conversion and adjustment.

To accelerate the load in a short period of time for a prime mover with limited output torque, the speed ratio *n of* its gearing mechanism is required to be optimal, i.e.

$$n = \sqrt{\frac{I_a}{I_m}} \tag{2.8}$$

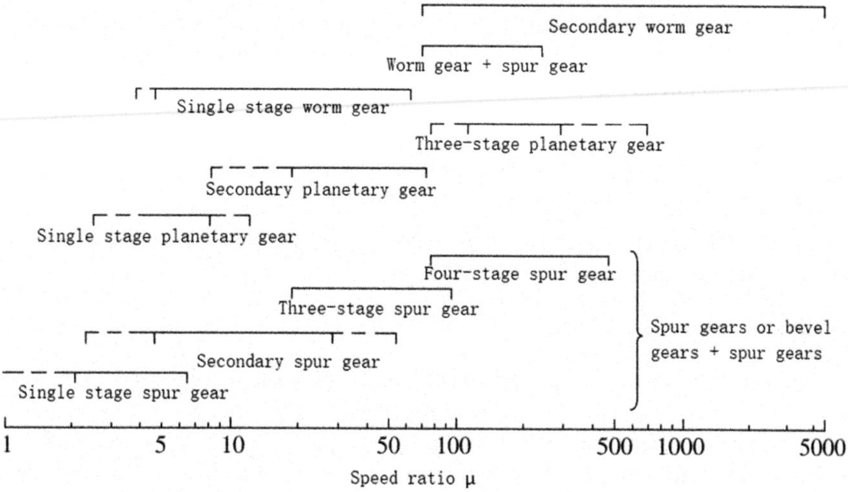

Fig. 2.13 Relationship between number of stages and speed ratio of gearing

where I_a is the moment of inertia of the working arm; I_m is the moment of inertia of the motor.

- Planetary Gear Reducer

Planetary gear reducers are broadly divided into S-C-P, 3S (3K), and 2S-C (2K-H) 3classes, with the structure shown in Fig. 2.14.

- S-C-P (K-H-V) type planetary gear reducer

The S-C-P consists of an internal gear, a planetary gear and a planetary gear holder. There is a certain offset between the center of the planetary gear and the center of the internal gear, and only some of the teeth participate in the engagement. The crank shaft is connected to the input shaft, and the planetary gear rotates around the internal gear while rotating on its axis. The number of revolutions of the planetary

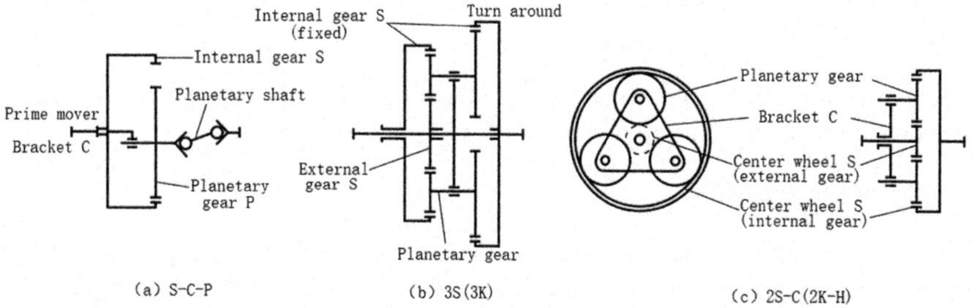

Fig. 2.14 Planetary gear reducer form

gear reverses while the planetary gear rotates in one week depends on the difference in the number of teeth between the planetary gear and the internal gear.

When the planetary gear is the output shaft, the transmission ratio is

$$i = \frac{Z_s - Z_p}{Z_p} \tag{2.9}$$

where Z_s is the number of teeth of the internal gear (sun gear); Z_p is the number of teeth of the planetary gear.

- 3S type planetary gear reducer

The planetary gears of the 3S type reducer mesh with two internal gears at the same time and also rotate around the sun gear (external gear). Of the two internal gears, the other gear can rotate when one is fixed and can be connected to the output shaft. The transmission ratio of this type of reducer depends on the difference in the number of teeth of the two internal gears.

- 2S-C type planetary gear reducer

The 2S-C type consists of two sun gears (outer and inner gears), planetary gears, and brackets. Two to four identical planetary gears are sandwiched between the inner and outer gears, and the planetary gears mesh with both the outer and inner gears. The bracket is connected to the center of each planetary gear, and the planetary gears force the bracket to rotate around the central wheel axis when they rotate in common.

In the above planetary gear mechanism, if the difference between the number of teeth Z_s *of the* internal gear and the number of teeth Z_p *of the* planetary gear is 1, the maximum reduction ratio $i = 1/Z_p$ can be obtained, but it is easy to produce mutual interference of the tooth tops, and this problem can be solved by the following method.

- Utilization of circular tooth shapes or steel balls.
- The number of tooth differences designed into 2.
- Planetary gears are made of thin elliptical shape that can be deformed elastically (harmonic drive).

- Harmonic Drive

The harmonic reducer is composed of the 3 basic parts of harmonic generator, flexible wheel and rigid wheel, as shown in Fig. 2.15.

- **Harmonic generator**

The harmonic generator is a component made of thin-walled bearings embedded in the outer circumference of an elliptical cam. The inner ring of the bearing is fixed to the cam and the outer ring is deformed elastically by steel balls, which are generally connected to the input shaft.

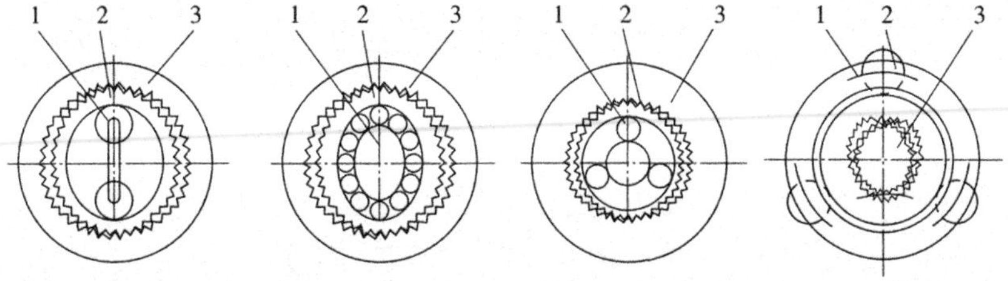

(a) Dual-wave contact type inner generator (b) Double-wave cam thin-wall bearing inner generator (c) Three-wave planetary inner generator (d) Three-wave planetary outer generator

1—harmonic generator; 2—flexible wheel; 3—rigid wheel

Fig. 2.15 Composition and type of harmonic drive mechanism

- **Soft Wheel**

The flex wheel is a cup-shaped thin-walled metal elastomer with teeth cut into the outer circle of the cup, and the bottom is called the bottom of the flex wheel, which is used to connect with the output shaft.

- **Rigid wheel**

The rigid wheel has many teeth in the inner circle, and the number of teeth is two more than that of the flexible wheel, which is usually fixed in the housing.

Harmonic generators are usually composed of cams or eccentrically mounted bearings. The rigid wheel is a rigid gear, and the flexible wheel is a gear that can produce elastic deformation. When the harmonic generator rotates continuously, the mechanical force generated deforms the flexible wheel in a process that forms a basically symmetrical harmony curve. The generator wave number indicates the number of cycles of deformation at a point of the flexible wheel when the generator rotates for one cycle. The working principle is that when the harmonic generator rotates inside the flexible wheel, it forces the flexible wheel to deform while entering or exiting between the teeth of the rigid wheel. In the direction of the short axis of the generator, the teeth of the rigid wheel and the flexible wheel are in the process of engaging or disengaging, and along with the continuous rotation of the generator, the meshing state of the teeth changes in turn, i.e., the process of engaging-meshing-engaging out-disengaging-engaging. This misaligned motion changes the input motion into the output deceleration motion.

The calculation of the harmonic drive speed ratio is the same as the calculation of the planetary drive speed ratio. If the rigid wheel is fixed, the harmonic generator ω_1 is the input and the flexible wheel ω_2 is the output, the speed ratio $i_{12} = \frac{\omega_1}{\omega_2} = -\frac{z_r}{z_g - z_r}$. If the flexible wheel is stationary, the harmonic generator w1 is the input, and the rigid wheel w3 is the output, the speed ratio $i_{13} = \frac{\omega_1}{\omega_3} = \frac{z_g}{z_g - z_r}$, where z_r is the number of teeth of the flexible wheel; z_g is the number of teeth of the rigid wheel.

The tooth circumference of flexible wheel and rigid wheel is equal, and the number of teeth is not equal, generally take the tooth difference of double wave generator is 2, and the tooth difference of three wave generator is 3. The stress generated by double

Fig. 2.16 Hydraulic hydrostatic wave generator harmonic transmission

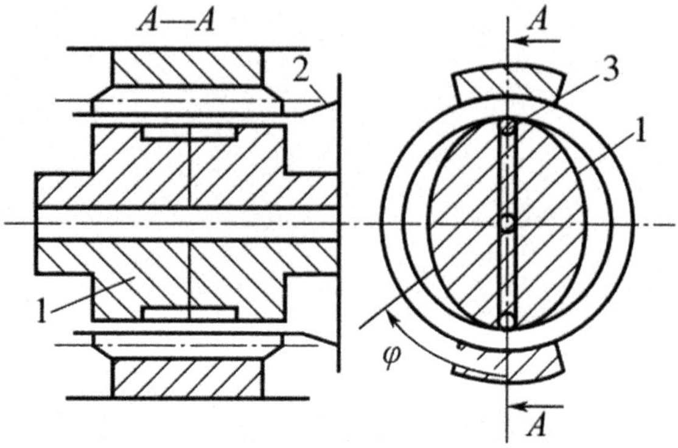

1—Cam; 2—Flexible wheel; 3—Small hole

wave generator when deforming the flexible wheel is small, and it is easy to obtain a large transmission ratio. The radial force required by the three-wave generator during the deformation of the flexible wheel is large, and the eccentricity of the transmission is small, which is suitable for precision indexing. The minimum number of teeth for harmonic transmission is usually recommended at $Z_{rmin} = 150$ for a tooth difference of 2 and $Z_{rmin} = 225$ for a tooth difference of 3.

Harmonic transmission is characterized by simple structure, small size, light weight, high transmission accuracy, large load capacity, large transmission ratio, and high damping characteristics. However, the flexible wheel is easy to fatigue, low torsional stiffness, and easy to generate vibration.

In addition, there is also a harmonic drive mechanism using a hydraulic hydrostatic wave generator and an electromagnetic wave generator. Figure 2.16 shows a schematic diagram of a harmonic drive using a hydraulic hydrostatic wave generator. There is no direct contact between the cam 1 and the flex wheel 2, and there is a gap of about 0.1 mm between the small hole 3 in the cam 1 and the inner surface of the flex wheel. High-pressure oil is ejected from the small hole 3 to produce a deformation wave in the flex wheel, thus realizing a deceleration-driven harmonic drive.

Harmonic drive mechanism has been widely used in robotics. The robot sent to the moon by the United States, the Rohren and Gerot R30 robots developed by Volkswagen in Germany and the Vertical 80 robot developed by Renault in France all use harmonic transmission mechanism.

2.4 Robot Arm

The robot arm is a component that supports the wrist and end-effector and is used to change the position of the end-effector in space. The structural form of the arm needs to be determined based on the robot's gripping weight, form of motion, positioning accuracy, degrees of freedom, and other factors. The main forms of motion of the robot arm are telescoping, pitching, slewing, lifting, etc., and the typical mechanisms to achieve its motion are.

- Telescopic motion mechanism

The telescopic movement of the robot arm causes the working length of its arm to change, and the common mechanisms to achieve its movement are piston hydraulic (pneumatic) cylinder, screw nut mechanism, piston cylinder and rack and pinion mechanism, piston cylinder and linkage mechanism, etc.

Piston hydraulic (pneumatic) cylinders are more frequently used in robot arm structures because of their small size and light weight. Figure 2.17 shows the telescopic structure of the double guide rod arm. The arm and wrist are mounted on the upper end of the lifting hydraulic cylinder through the connecting plate. When the two chambers of the double-acting hydraulic cylinder 1 are respectively fed with pressure oil, the piston rod 2 (i.e., the arm) is pushed to make reciprocating linear motion. As the arm of the telescopic hydraulic cylinder is installed between the two guide rods, the guide rod to bear the bending role, the piston rod only by the pulling role, so the force is simple, smooth transmission, neat and beautiful appearance, compact structure. Figure 2.18 is the arm telescopic structure using four guide columns. The vertical telescopic movement of the arm is driven by the cylinder 3. It is characterized by long stroke and heavy grip. When the shape of the workpiece is irregular, four guide pillars are used to prevent large bias moment.

- **Pitching motion mechanism**

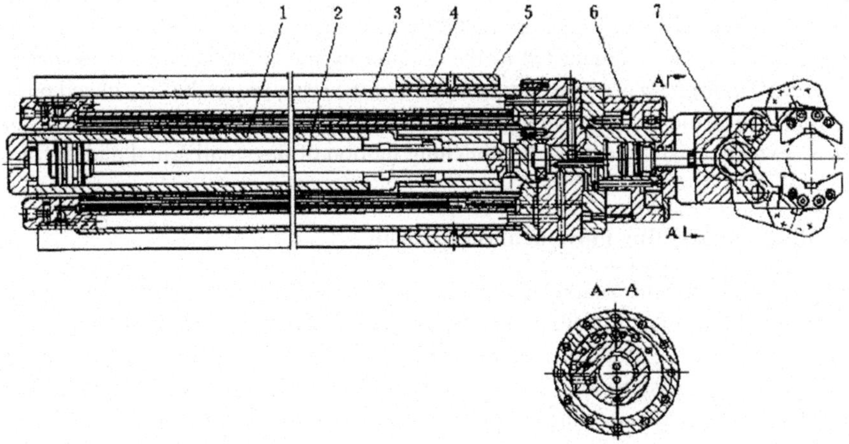

1. double acting hydraulic cylinder; 2. piston rod; 3. guide rod;
4. guide sleeve; 5. support seat; 6. wrist; 7. hand

Fig. 2.17 Telescopic structure of double guide bar arm

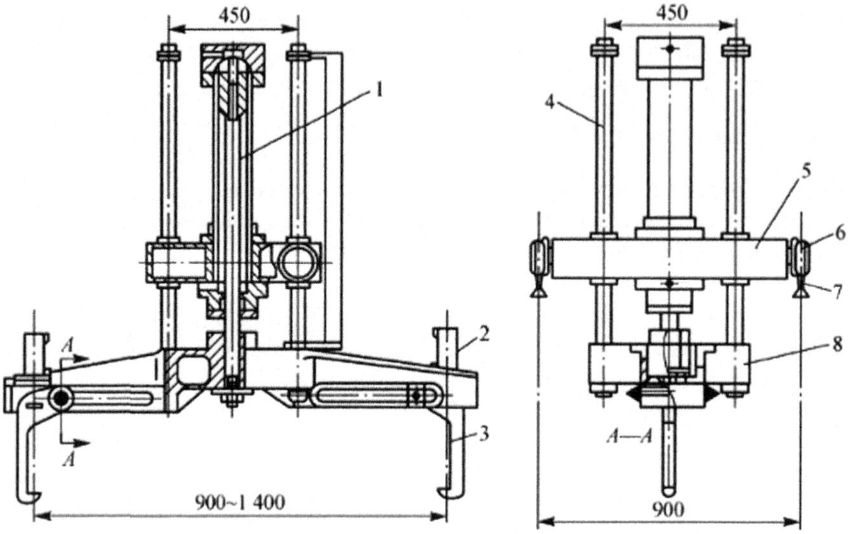

1. Hand; 2.Clamping cylinder; 3. Oil cylinder; 4. Guide column; 5. Running frame; 6.Traveling wheels; 7.Rail; 8. Support

Fig. 2.18 Four guide column type arm telescoping mechanism

The arm pitch motion of the robot is generally realized by using a piston hydraulic cylinder and a linkage mechanism. Figure 2.19 shows the schematic diagram of the arm pitch drive cylinder installation. The piston cylinder for the arm pitch motion is located under the arm, and its piston rod and arm are connected by hinge, and the cylinder body is connected to the column by means of tail earring or central pin, etc. Figure 2.20 is a schematic diagram of the structure of the articulated piston cylinder to achieve arm pitching. It uses the articulated piston cylinder 5, 7 and linkage mechanism to achieve the pitching motion of the small arm 4 and the large arm 6.

- **Rotary and lifting movement mechanism**

Slewing motion is the rotation of the robot around the lead hammer axis. This motion determines the angular position that the robot's arm can reach. The common mechanisms to realize the lifting and slewing motion of the robot arm are vane type slewing cylinder, gear transmission mechanism, sprocket transmission mechanism, linkage mechanism, etc. The slewing cylinder is driven separately from the lifting cylinder and is suitable for cases where the lifting stroke is short and the slewing angle is less than 360°. Figure 2.21 is the use of lift cylinder and rack and pinion drive structure to achieve the arm lift and rotation movement of the schematic diagram, the rack and pinion mechanism is through the reciprocating motion of the rack, drive and arm connected to the gear to do reciprocating rotary motion, so as to achieve the

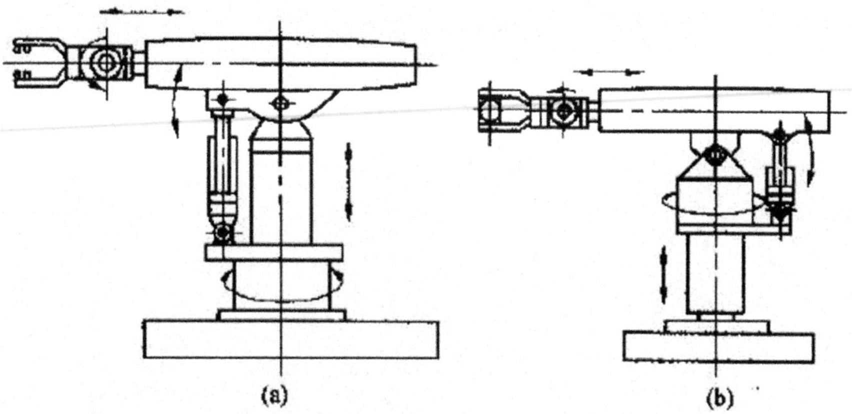

(a) Schematic diagram 1 (b) Schematic diagram 2

Fig. 2.19 Schematic diagram of arm pitching drive cylinder installation

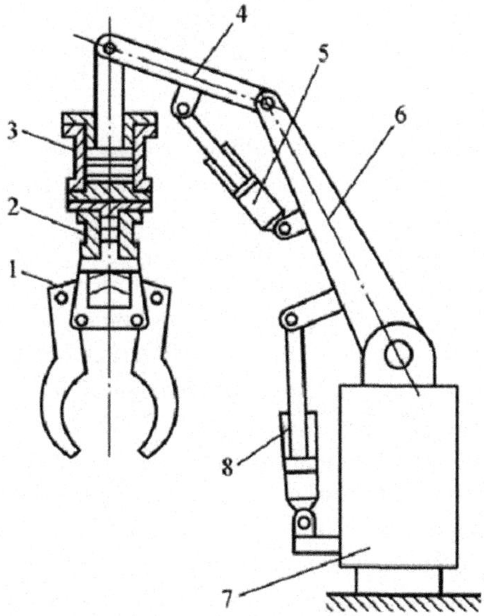

1. Arm; 2.Clamping cylinder; 3.Lifting cylinder; 4.Small arm; 5, 7.Articulated piston cylinder; 6.Large arm; 8.Column

Fig. 2.20 Schematic diagram of the structure of the articulated piston cylinder for arm pitching

Fig. 2.21 Structure of arm lifting and slewing motion

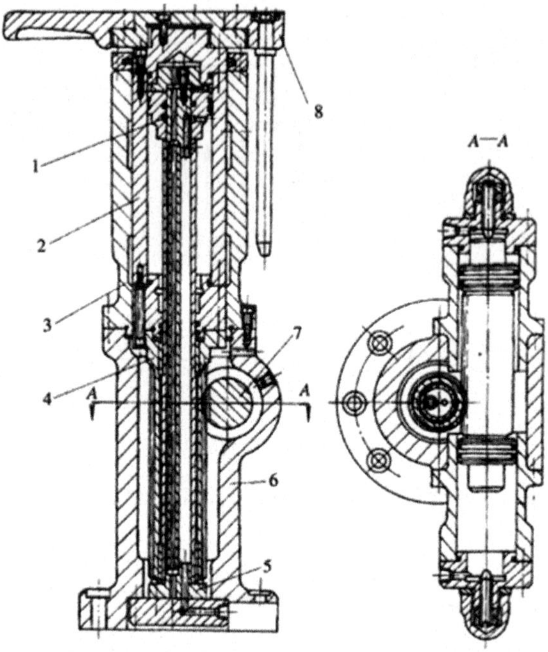

1. Piston rod; 2.Lift cylinder body; 3.Guidance sleeve; 4.Gear; 5.Connection cover; 6.Base; 7.Rack; 8.Connection plate

rotary motion of the arm. The piston cylinder that drives the reciprocating movement of the rack can be driven by pressure oil or compressed gas. Piston hydraulic cylinder two cavities into the pressure oil, push the rack piston 7 to do reciprocating movement (see A-A section), and the rack 7 meshing gear 4 that do reciprocating rotary motion. As the gear 4, arm lifting cylinder 2, connecting plate 8 are connected with screws into one, connecting plate and arm solid connection, so as to achieve the rotary motion of the arm. The piston rod of the lift hydraulic cylinder is connected to the seat 6 through the connection cover 5 and fixed, the cylinder body 2 along the guide sleeve 3 to do up and down, because the external lift hydraulic cylinder is equipped with guide sleeve, so rigidity is good, transmission is smooth.

2.5 Robot Dexterity Hand

Robotics has developed to the stage of intelligence, and the robot's hands have evolved from industrial robots used to carry objects, assemble parts, weld, and spray paint, to become increasingly dexterous, capable of fine and complex tasks such as undersea rescue, holding pens and writing, playing musical instruments, and grabbing eggs.

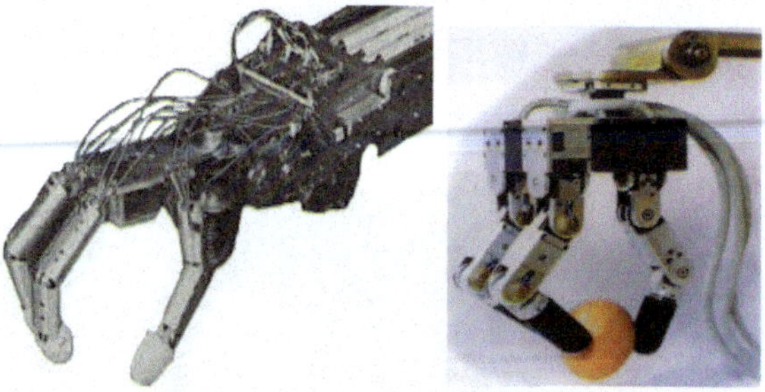

Fig. 2.22 Three-finger dexterity hand

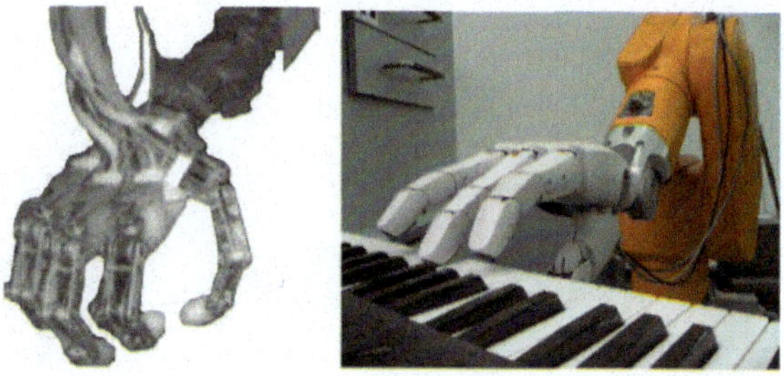

Fig. 2.23 Four-finger dexterity hand

The robotic multi-finger dexterous hand that mimics the human hand can improve the manipulation ability, dexterity and rapid response of the robot hand, enabling the robot to perform various complex operations like the human hand. The multi-finger dexterous hand has multiple fingers, each with three rotary joints, and the degrees of freedom of each joint are independently controlled. Therefore, it can imitate human fingers to perform various complex movements, such as writing and playing piano. Figures 2.22, 2.23 and 2.24 show the three-finger, four-finger and five-finger dexterous hands, respectively. Usually, the multi-finger dexterous hand is equipped with force, vision, touch, temperature and other sensors, which can be applied not only in the field of grasping various shaped objects and performing various humanoid operations, but also in the field of completing operations that cannot be achieved by humans under various extreme environments, such as space, disaster rescue and so on. Figure 2.25 gives examples of various applications of five-finger dexterous hands, which can grasp various shaped objects and perform various humanoid operations. Figure 2.26 presents an example of the integration of the dexterous hand with 3D vision sensors and force sensors for applications.

2.6 Common Moving Mechanism

The main forms of mobile mechanisms for mobile robots are: wheeled mobile mechanism; crawler mobile mechanism; legged mobile mechanism. In addition, there are also stepping type mobile mechanism, creeping type mobile mechanism, hybrid type

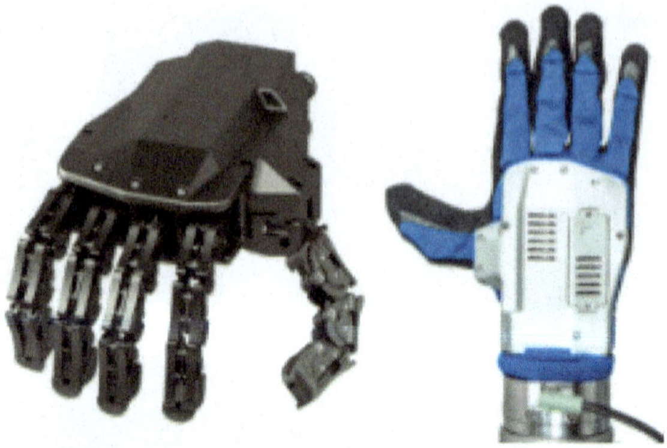

Fig. 2.24 Five-finger dexterity hand

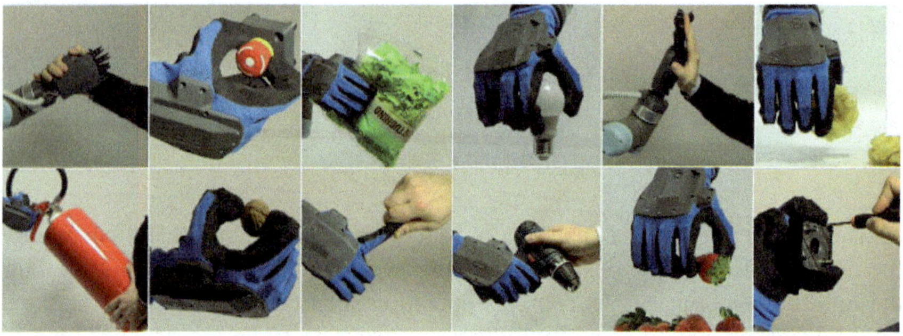

Fig. 2.25 Example of the main application scenarios of the five-finger dexterous hand

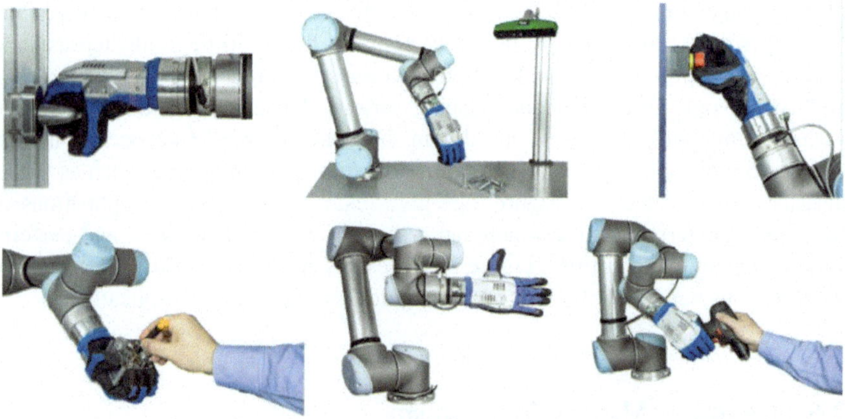

Fig. 2.26 Dexterous hand with 3D vision sensors and force sensors for integrated applications

mobile mechanism and snake type mobile mechanism, which are suitable for various special occasions.

2.6.1 Wheel Type Moving Mechanism

Wheel type mobile mechanism can be classified by the number of wheels.

- Two-wheeled vehicles

Experiments in which people used very simple and inexpensive bicycles or motorcycles for robotics were conducted early on. But it was easily recognized that the physical quantities such as speed and tilt of two- wheeled vehicles were not very accurate, and simple, inexpensive and reliable sensors required for robotization were difficult to obtain. In addition, the two-wheeled vehicle is extremely unstable when braking and when walking at low speeds. Figure 2.27 shows a motorcycle equipped with a gyroscope. When people drive a two-wheeled vehicle, they rely on hand operation and the movement of the center of gravity in order to drive stably. This gyroscopic two-wheeled vehicle, which applies a torque proportional to the tilt of the vehicle to the axle system, uses the gyroscopic effect to make the vehicle stable.

- Tricycle

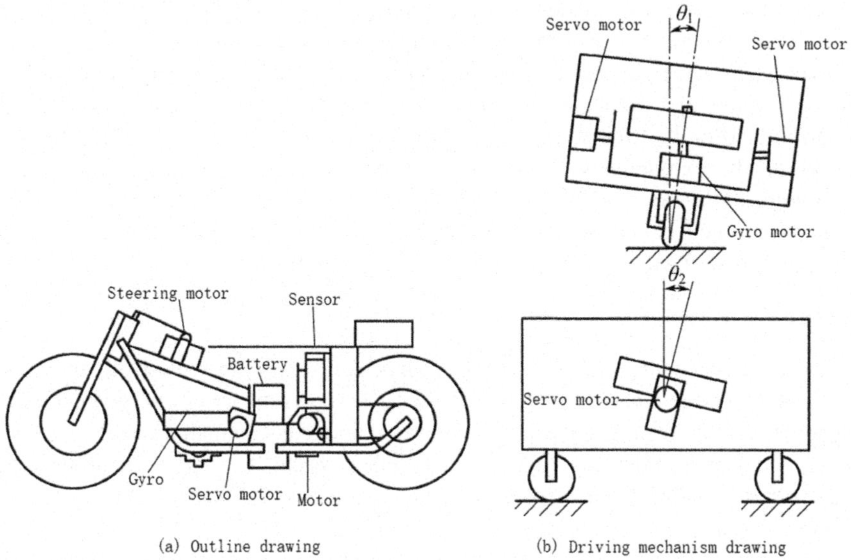

Fig. 2.27 Two-wheeled vehicle using gyroscope

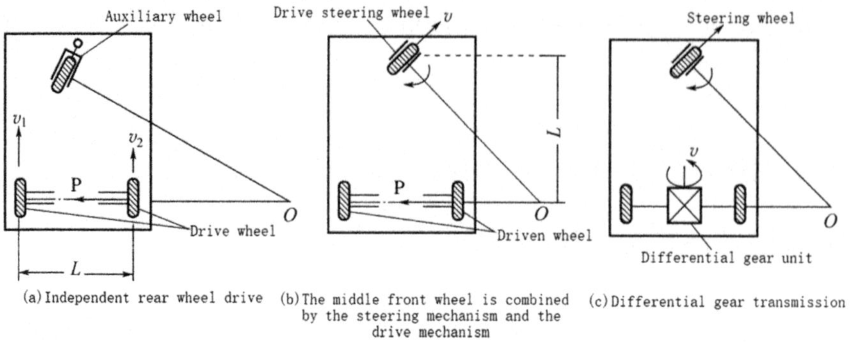

(a) Independent drive of the rear wheels; (b) Combination of the middle and front wheels by the steering mechanism and the driving mechanism; (c) Differential gear drive

Fig. 2.28 Mechanism of tricycle-type mobile robot

The three-wheel moving mechanism is the basic moving mechanism of the wheel-based robot, and its principle is shown in Fig. 2.28.

Figure 2.28a is a combination of the rear wheel with two wheels driven independently and the front wheel with an auxiliary wheel composed of small casters. This mechanism is characterized by the simple composition of the mechanism, and the radius of rotation can be set arbitrarily from 0 to infinity. However, its center of rotation is on the line connecting the two driving axes, so the center of rotation is not consistent with the center of the car even if the radius of rotation is 0.

The front wheel in Fig. 2.28b consists of a combined rudder mechanism and a drive mechanism. Compared with Fig. 2.28a, the drives of both rudder and drive are concentrated in the front wheel part, so the mechanism is complex and its rotation radius can be continuously varied from 0 to infinity.

Figure 2.28c shows the method of driving by differential gears to avoid the disadvantages of the mechanism in Fig. 2.28b. Recently, the differential gear is no longer used, but the left and right wheels are driven separately and independently.

- Four-wheelers

The driving mechanism and motion of a four-wheeled vehicle are basically the same as those of a three-wheeled vehicle. Figure 2.29a shows the way of two-wheel independent drive with auxiliary wheels at the front and rear. Compared with Fig. 2.28a, it is good for changing direction in narrow places because it can rotate around the center of the vehicle when the rotation radius is 0. Figure 2.29b is the car way, which is suitable for high-speed travel and good stability.

Depending on the purpose of use, there are also vehicles using six-wheel drive and tire vehicles with different wheel diameters, and there are also proposals to use vehicles with flexible mechanisms. Figure 2.30 shows an example of a small rover for Mars exploration, which has wheels that can adjust their height up and down

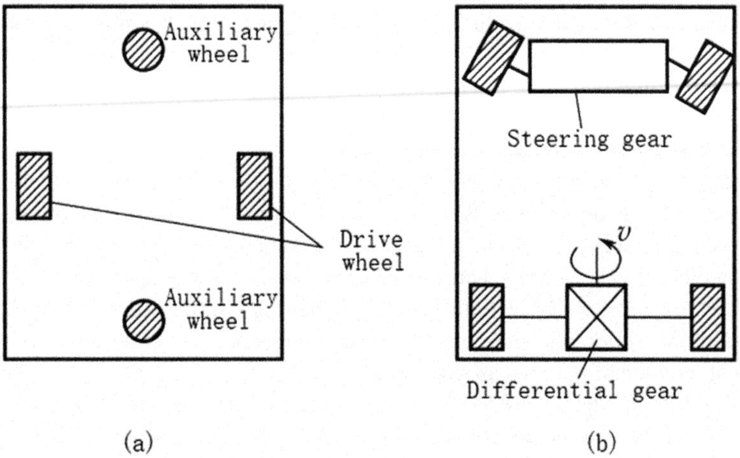

Fig. 2.29 Drive mechanism and motion of a four-wheeled vehicle

according to the terrain to improve their stability and suitability for operation on the surface of Mars.

- Omni-directional mobile car

The previous wheel-based mobile mechanism is basically two degrees of freedom, so it is not possible to simply achieve arbitrary positioning and orientation of the car. Orientation of the robot with a four-wheeled composition of the car can be achieved by controlling the steering angle of each wheel. Omni-directional moving

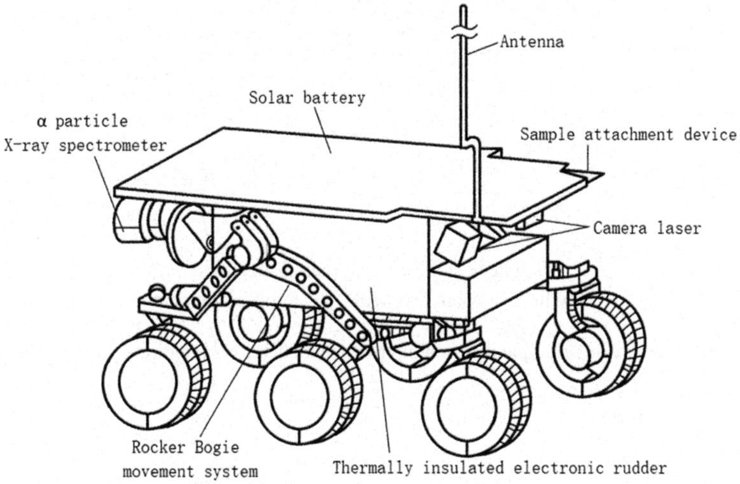

Fig. 2.30 Small Rover for Mars exploration

mechanism can move in any direction along the plane while keeping the orientation of the body unchanged. Some Omni-directional wheel mechanisms can change the orientation of the body like an ordinary vehicle, in addition to having the ability to move in all directions. Due to the flexible handling performance of this mechanism, it is particularly suitable for mobile operations in narrow spaces (passages).

Figure 2.31 is a transmission principle diagram of an all-wheel deflection type all-round moving mechanism. When the travel motor M_1 is running from, the wheel 1 is driven to rotate by the worm-worm gear pair 5 and the bevel gear pair 2. When the steering motor M_2 is running. Through another pair of worm gears 6 and gears 3 drive the wheel bracket 4 to deflect appropriately. When each wheel takes different combinations of deflection and the corresponding wheel speed, it is possible to achieve different movement patterns as shown in Fig. 2.32.

A new type of wheel called Mecanum wheels is used in the more widely used all-round four-wheel moving mechanism. Figure 2.33a shows the shape of the Mecanum wheel, which consists of two parts, namely the active hub and a number of passive rollers evenly distributed along the outer edge of the hub in a certain direction. When the wheel rotates, the velocity v of the wheel core relative to the ground is a synthesis of the hub velocity v_h and the roller rolling velocity V_r. The v_h has a deviation angle θ from v, as shown in Fig. 2.33b. Since each wheel has this feature, the all-round movement and in-situ steering motion of the vehicle can be realized after proper combination, see Fig. 2.34.

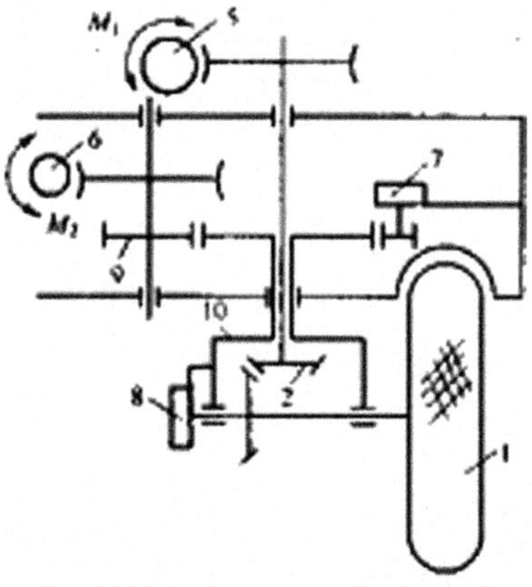

1. wheel; 2. bevel gear pair; 3. gear pair;
4.wheel bracket; 5,6. worm gear and worm pair

Fig. 2.31 Transmission principle diagram of all-wheel deflection omnidirectional moving mechanism

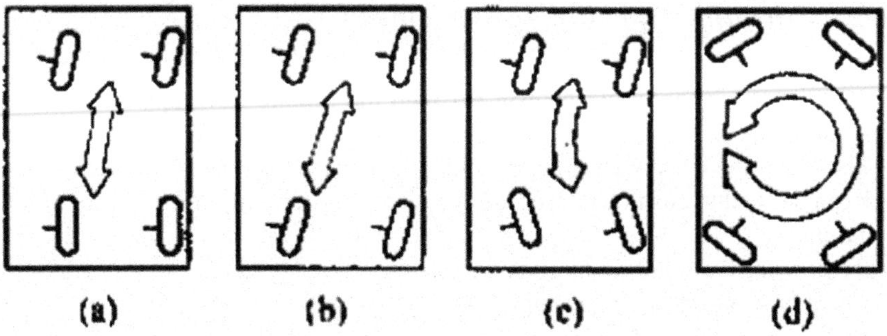

(a) front-wheel steering; (b) all-round approach; (c) four-wheel steering; (d) in-situ slewing

Fig. 2.32 The movement mode of the all-wheel deflection omnidirectional vehicle

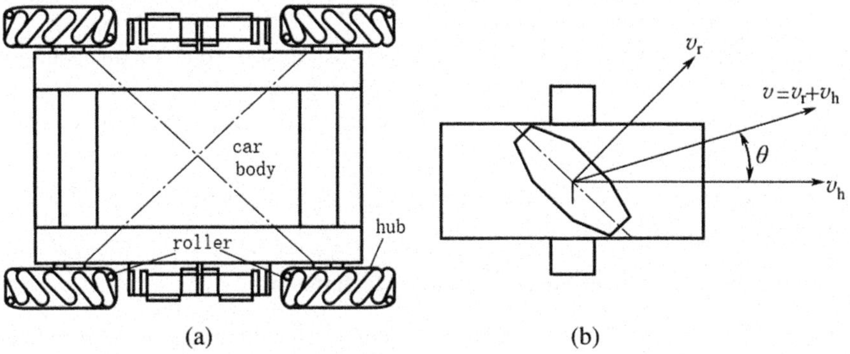

Fig. 2.33 McCannum wheel and its speed synthesis

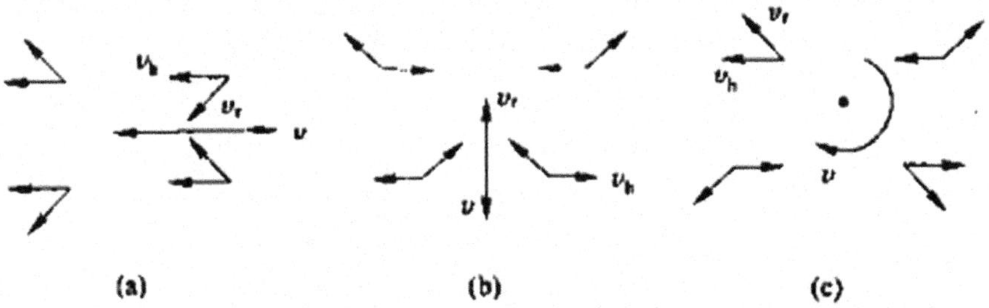

(a) Longitudinal; (b) Transverse; (c) Steering

Fig. 2.34 Speed configuration and movement of McCannam vehicles

2.6.2 Crawling Type Mobile Mechanism

The crawler type mobile mechanism is called the infinite track approach, and its most important feature is that the circular infinite tracks are wound around multiple wheels so that the wheels do not come into direct contact with the road surface. The track can be used to cushion the state of the road, so it can travel on a variety of road conditions.

The crawler type mobile mechanism has the following features compared with the wheel type mobile mechanism.

- Large supporting area, small grounding specific pressure. Suitable for soft or muddy field for operation, small sinkage, small rolling resistance, better passing performance.
- Good off-road mobility, climbing, crossing ditch and other performance are better than wheeled mobile agencies.
- Track teeth on the track support surface, not easy to slip, good traction adhesion performance, conducive to play a larger traction force.
- Complex structure, high weight, high motion inertia, poor vibration damping performance, and easily damaged parts.

Common track drive mechanisms include tractors, tanks, etc. Several special track structures are described here.

- **Caterpillar Overhead Sprocket Track Mechanism**

The overhead sprocket track mechanism is a non-equilateral triangular shaped track mechanism developed by Caterpillar, USA, with the drive wheels elevated and a semi-rigid suspension or elastic suspension, as shown in Fig. 2.35.

Compared with the conventional crawler travel mechanism, the overhead sprocket flexible suspension travel mechanism has the following features.

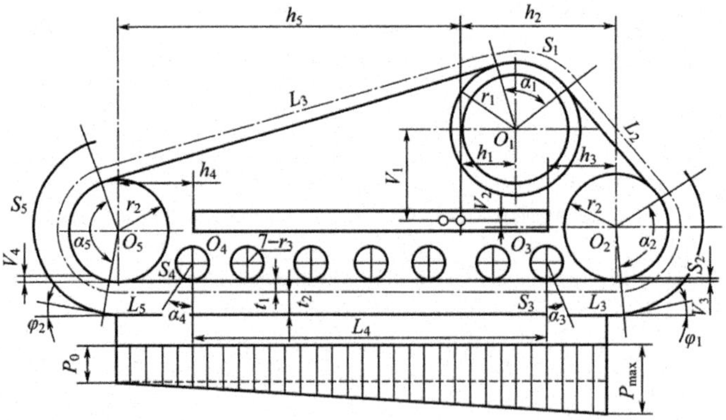

Fig. 2.35 Schematic diagram of overhead sprocket track movement mechanism

- By placing the drive wheel high, not only is the load from outside isolated, but all the load is absorbed by the suspended oscillating mechanism and pivot and not directly transmitted to the drive sprocket. The drive sprocket is only subjected to torsional loads and is kept away from the ground environment, reducing the wear between the sprocket teeth and links caused by debris being carried in.
- The elastic suspension travel mechanism can keep more tracks in contact with the ground to distribute the load evenly. Therefore, smaller size parts can be used for the same machine weight.
- The elastic suspension running mechanism has the advantages of large load capacity, smooth running, low noise, large ground clearance and good adhesion, which makes the machine have higher motor flexibility without sacrificing stability and reduces the power loss due to track slippage.
- Easy maintenance of each part of the travel mechanism.

• Variable shape track mechanism

Shape-variable track mechanism refers to a mechanism in which the configuration of the track can be changed as needed. Figure 2.36 shows the shape of a variable shape track. It consists of two tracks with variable shape, which are driven by two main motors. When the two tracks have the same speed, forward or backward movement is achieved; when the two tracks have different speeds, the whole machine achieves steering movement. When the main arm bar rotates around the shaft on the track frame, it drives the planetary wheels to rotate, thus realizing different shapes of the tracks to adapt to different moving environments.

• Position variable track mechanism

Position-variable track mechanism refers to a track mechanism in which the position of the track relative to the body can be changed. This change in position can be in one or two degrees of freedom. Figure 2.37shows a two-degree-of-freedom variable

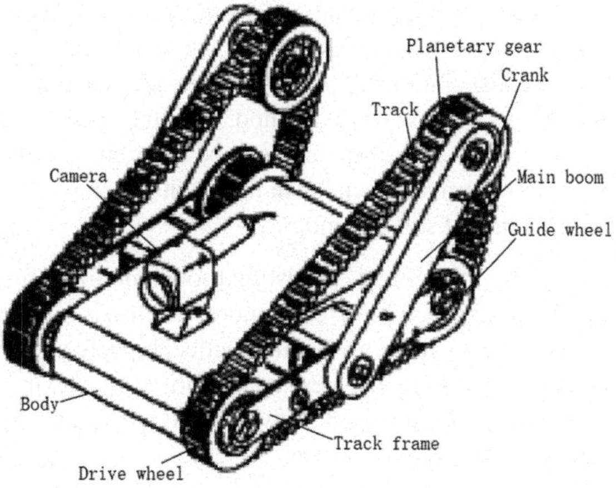

Fig. 2.36 Shape-variable track moving mechanism

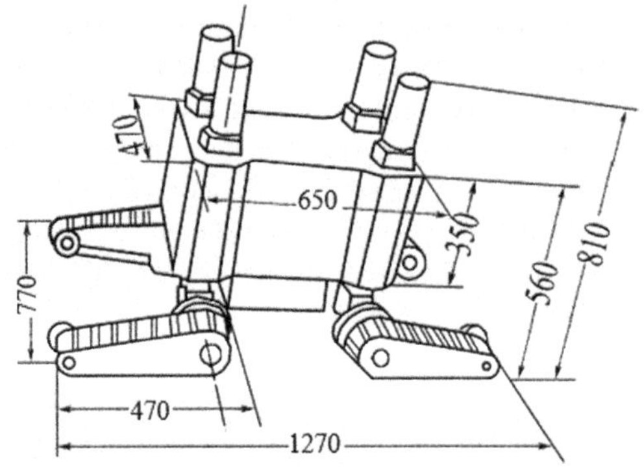

Fig. 2.37 Two-degree-of-freedom variable displacement track moving mechanism

position track movement mechanism. Each track is capable of deflecting around the horizontal and vertical axes of the body, thus changing the overall configuration of the moving mechanism. When the tracks are shifted along one degree of freedom, they are used for climbing over steps and crossing ditches; when they are shifted along the other degree of freedom, they can be used in an all-round way on wheels.

2.6.3 *Leg and Foot Moving Mechanism*

Although a tracked mobile mechanism can move on uneven ground, it is not very adaptable, has a large sway when walking and is inefficient when driving on soft ground. According to the survey, nearly half of the earth's ground is not suitable for traditional wheeled or tracked vehicles to walk. However, the average multilegged animal is able to move freely in these places, and it is clear that footed mobility mechanisms have unique advantages in such environments.

- The foot-mobile mechanism is well adapted to rough terrain. The foot-mobile mode has discrete points of foothold and can choose the optimal support point on the ground that can be reached, while the wheeled and tracked mobile mechanism must face almost all points on the worst terrain.
- The foot movement method also has active vibration isolation, so that the movement of the airframe can be quite smooth despite the uneven ground level.
- The foot travel mechanism moves at a higher speed and consumes less energy on uneven and soft ground.

The number of feet of existing footed mobile robots are single, bipedal, triple and quadruple, six, eight or even more. The high number of feet is suitable for heavy loads and slow movements. In practice, bipedal and quadrupedal are used most often because they have the best adaptability and flexibility and are closest to humans and animals. Figure 2.38 shows ASIMO, a humanoid robot developed in Japan, and Fig. 2.39 shows a robot dog.

Fig. 2.38 Humanoid robot ASIMO

Fig. 2.39 The robot dog

2.6.4 Other Forms of Moving Mechanism

For special purposes, various kinds of moving mechanisms have also been developed, such as the adsorption moving mechanism on the wall, the snake mechanism, etc. Figure 2.40 shows a robot that can crawl on a wall, where (a) the figure is a suction cup interactively adsorbed on the wall to move, and (b) the roller shown in the figure is a magnet, and the wall must be a magnetic material to work. Figure 2.41 shows the snake robot.

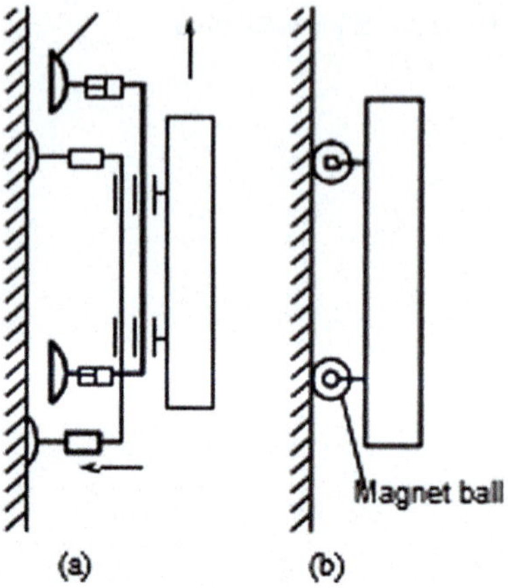

Fig. 2.40 Wall climbing robot

Fig. 2.41 Snake robot

Chapter 3
Sensing Technology for Special Robots

3.1 Overview

The robot sensor can be defined as a device that can transform the characteristics (or parameters) of the robot target into electrical output. The robot achieves a human-like perception through the sensor, which is called the robot's electrical facial features.

As an important industry, the development of robots is in the ascendant, and its application range is increasingly extensive. It is required that it can perform more and more complex work, have stronger adaptability to changing environments, and require more precise positioning and control. The application is not only very necessary, but also has higher requirements.

3.1.1 Sensor Requirements for Special Robots

- Basic performance requirements

– High precision and good repeatability

The precision of the robot sensor directly affects the work quality of the robot. The sensors used to detect and control the motion of the robot are the basis for controlling the positioning precision of the robot. Whether the robot can work unmistakably often depends on the measurement precision of the sensors.

– Good stability and high reliability

The stability and reliability of robot sensors are necessary conditions to ensure that the robot can work stably and reliably for a long time. Robots are often operated in

The original version of this chapter was revised: Figure 3.26 has been corrected. The correction to this chapter is available at https://doi.org/10.1007/978-981-99-0589-8_8

T. Guo et al., *Special Robot Technology*, Advanced and Intelligent Manufacturing in China, https://doi.org/10.1007/978-981-99-0589-8_3

place of humans under unattended conditions. If it fails at work, the light ones it will affect the normal production, and cause serious accidents in severe cases.

– Strong anti-interference ability

The working environment of the robot sensor is relatively harsh, it should be able to withstand strong electromagnetic interference, strong vibration, and can work normally in a certain high temperature, high pressure and high pollution environment.

– Small quality, small size, convenient and reliable installation

For sensors installed on moving parts such as robot manipulators, the weight should be small, otherwise the inertia of the moving parts will increase and the motion performance of the robot will be affected. For robots whose workspace is somewhat restricted, requirements for volume and installation direction are also essential.

- Job requirements

The environmental perception ability is the most basic ability of a mobile robot besides moving. The level of perception ability directly determines the intelligence of the mobile robot, and the perception ability is determined by the perception system. The perception system of a mobile robot is equivalent to the five sense organs and nervous system. It is a tool for the robot to obtain external environmental information and perform internal feedback control. It is one of the most important parts of a mobile robot. The perception system of a mobile robot is usually composed of a variety of sensors, which are located at the interface between the external environment and the mobile robot, and are the window for the robot to obtain information. The robot uses these sensors to collect various information, and then takes appropriate methods to comprehensively process the environmental information obtained by multiple sensors to control the robot to perform intelligent operations.

3.1.2 Characteristics of Common Sensors

When selecting the suitable sensor to suit a specific need, many different characteristics of the sensor must be considered. These characteristics determine the performance of the sensor, whether it is economical, whether it is easy to use, and the scope of application. In some cases, different types of sensors can be selected to achieve the same goal. The following factors should generally be considered before selecting a sensor.

- Cost

The cost of sensors is an important factor to consider, especially when multiple sensors are used on a machine. However cost must be balanced against other design requirements such as reliability, importance of sensor data, precision and longevity.

- Size

Depending on the application of the sensor, size may sometimes be the most important. For example, joint displacement sensors must fit with the design of the joint and be able to move with other components in the robot, but the space available around the joint may be limited. Additionally, huge sensors can limit the range of motion of the joints. Therefore, it is very important to make sure that there is enough space for the joint sensors.

- Weight

Since the robot is a moving device, the weight of the sensor is important, too much sensor weight will increase the inertia of the manipulator, while also reducing the overall payload.

- Type of output (digital or analog)

Depending on the application, the output of the sensor can be digital or analog, and they can be used directly, or they may have to be converted before they can be used. For example, the output of a potentiometer is analog, while the output of an encoder is digital. If an encoder is used with a microprocessor, its output can be passed directly to the input of the processor, while the output of the potentiometer must be converted to a digital signal using an analog-to-digital converter (ADC). Which output type is more suitable must be considered in combination with other requirements.

- Interface

Sensors must be able to interface with other devices such as microprocessors and controllers. If the interface between the sensor and other devices does not match or other additional circuits are required between the two, the interface between the sensor and the device needs to be solved.

- Resolution ratio

Resolution ratio is the minimum value the sensor can resolve within the measurement range. In a wire wound potentiometer, it is equivalent to one turn of resistance. In an n-bit digital device, resolution ratio = full scale/2n. For example, a four-bit absolute encoder can have up to 24 = 16 different levels when measuring position. Therefore, the resolution ratio is 360°/16 = 22.5°.

- Sensitivity

Sensitivity is the ratio of the output response change to the input change. The output of a high-sensitivity sensor will have large fluctuations due to input fluctuations (including noise).

- Linearity

Linearity reflects the relationship between input variables and output variables. This means that a sensor with a linear output will produce the same change in output for any same change in input over its range. Almost all devices have some nonlinearity in nature, but the degree of nonlinearity varies. Within a certain operating range, some devices can be considered linear, while others can be linearized with certain preconditions. If the output is not linear, but the nonlinearity is known, the nonlinearity can be overcome by appropriately modeling it, adding measurement equations, or additional electronics. For example, if the output of a displacement sensor varies as the sine of the angle, then when applying this type of sensor, the designer can scale the output by the sine of the angle, either through an application program, or a simple circuit that can scale the signal by the sine of the angle.

- Range

Range is the difference value between the maximum and minimum output that the sensor can produce, or the difference value between the maximum and minimum input when the sensor is operating normally.

- Response time

Response time is the time it takes for the output of a sensor to reach a certain percentage of the total change, and it is usually expressed as a percentage of the total change, such as 95%. Response time is also defined as the time it takes to observe a change in the output when the input changes. For example, a simple mercury thermometer has a long response time, while a digital thermometer based on radiant heat has a short response time.

- Frequency response

If you connect small, cheap speakers to a high-performance radio, the speakers will restore the sound, but the sound quality will be poor, and a high-quality speaker system with both bass and treble will have good sound quality when restoring the same signal. This is because the frequency response of a two-horn speaker system is very different from that of a small, inexpensive speaker. Because the natural frequencies of a small speaker are higher, it can only reproduce higher frequency sounds. A speaker system with at least two speakers can restore the sound signal in the high and low speakers. One of the two speakers has a high natural frequency and the other has a low natural frequency. The fusion of the two frequency responses allows the speaker system to reproduce a very good sounding signal (in fact, the signal is filtered before going to the speakers). All systems can syntony around their natural frequencies with just a small excitation. The response diminishes as the excitation frequency decreases or increases. The frequency response bandwidth specifies the range within which the system responds relatively well to the input. The larger the bandwidth of the frequency response, the better the ability of the system to respond to different inputs. It is important to consider the frequency response of the sensor and determine if the sensor responds fast enough under all operating conditions.

- Reliability

Reliability is the ratio of the normal operation times of the system to the total operation times. For the situation requiring continuous operation, while considering the cost and other requirements, a reliable and long-term continuous operation sensor must be selected.

- Precision

Precision is defined as how close the sensor's output value is to the expected value. A sensor has an expected output for a given input, and precision is related to how close the sensor's output is to that expected value.

- Repeating precision

For the same input, if the output of the sensor is measured multiple times, the output may be different each time. Repeating precision reflects the degree of variation between multiple outputs of the sensor. Usually, if a sufficient number of measurements are made, a range can be determined that includes all measurements around the nominal value, and this range is defined as repeatability. Often Repeating precision is more important than precision, and in most cases inaccuracies are caused by systematic errors which can be corrected and compensated for because they can be predicted and measured. Repeatability errors are often random and cannot be easily compensated for.

3.1.3 Classification of Robot Sensors

Robots are equipped with different types and specifications of sensors according to different tasks. They are generally divided into internal sensors and external sensors. With the development of science and technology, the current sensor technology is also constantly developing, smart sensors have emerged, and wireless sensor network technology has also been developed rapidly.

The so-called internal sensor is a functional component that measures the state of the robot itself. The specific detected objects include geometric quantities such as linear and angular displacements with joints, motion quantities such as velocity, angular velocity, acceleration, and physical quantities such as inclination angle, azimuth angle, and vibration. It is mainly used to collect information from the inside of the robot. Table 3.1 lists the basic forms of sensors inside the robot.

The so-called external sensors are mainly used to collect the information of the interaction between the robot and the external environment and the work object,

Table 3.1 Basic classification of sensors inside robots

Internal sensor	Basic type
Position sensor	Potentiometer, rotary transformer code board
Speed sensor	Tachometer generator code board
Accelerometer sensor	Strain type, servo type, piezoelectric type, electric type
Tilt angle sensor	Liquid type, vertical oscillator type
Force (torque) sensor	Strain type, piezoelectric type

so that the robot and the environment can interact, so that the robot has the ability to self-correct and self-adapt to the environment. Robot external sensors usually include force, touch, vision, hearing, smell and proximity sensors. Table 3.2 lists the detection content and applications of these sensors.

Internal sensors and external sensors are divided according to the role of sensors in the system, some sensors can be used as both internal sensors and external sensors. For example, a force sensor, used in end effector or arm self-weight compensation, is an internal sensor; when measuring the reactive force of an operating object or obstacle, it is an external sensor.

Table 3.2 Basic classification of robots external sensors

Sensor	Test content	Detection device	Application
Force sense	Grasp Load Distributed pressure Moment Multiple forces Slide	Strain gauges, semiconductor pressure-sensitive components Spring displacement gauge Conductive rubber, Pressure sensitive polymer material Piezoresistive elements, motor galvanometers Strain gauges, semiconductor pressure-sensitive components Optical rotation detectors, optical fibers	Grasp control Tension control, finger pressure control Posture and shape discrimination Coordinated control Assembly force control Slip judgment, force control
Touch	Get in touch with	Limit switch	Action sequence control
Visual	Plane position Shape Distance Defect	ITV cameras, position sensors Line image sensor Rangefinder Area image sensor	Position determination, control Object recognition and discrimination Mobile control Inspection, anomaly detection
Hearing	Sound Ultrasound	Microphone Ultrasonic sensor	Language control (man–machine interface) Mobile control
Smell	Gas composition	Gas sensor, Ray sensor	Chemical composition detection
Closeness	Near Interval Tilt	Photoelectric switch, LED, Laser, infrared Phototransistors, photodiodes Electromagnetic coils, Ultrasonic sensors	Action sequence control Obstacle avoidance Trajectory movement control, exploration

3.2 Force Sensor

Force sense refers to the perception of the forces exerted on the robot's fingers, limbs and joints, mainly including wrist force sense, joint force sense and bearing force sense. According to the load of the measured object, the force sensor can be divided into a load cell (uniaxial Force Sensor), a torque meter (uniaxial torque sensor), a finger sensor (an ultra-small uniaxial force sensor that detects the force of the robot finger) and Six-axis force sensor, etc.

- Cross wrist force sensor

Figure 3.1 shows a flexible cross-beam wrist force sensor, which is cut into a cross frame from aluminum, the outer end of each cantilever beam is inserted into the inner hole of the circular wrist frame, and the joint between the end of the cantilever beam and the wrist frame is equipped with nylon balls, the purpose is to make the cantilever beam easy to expand. In addition, in order to increase its sensitivity, a slit is also cut in the wrist frame where it meets the beam. The cross-shaped cantilever beam is actually a whole, and its center is fixed in the axial direction of the wrist.
The strain gauges are attached to the cross beams, and one strain gauge is attached to the upper, lower, left, and right sides of each beam. Two strain gauges on the opposite side form a set of half bridges, and a parameter can be detected by measuring the output of one half bridge. The entire wrist can detect 8 parameters through the strain gauge: fx1, fx3, fy1, fy2, fy3, fy4, fz2, fz4, and these parameters can be used to calculate the force Fx, Fy, Fz and the torques Mx, My, Mz, in the x, y and z directions are shown in Formula (3.1).

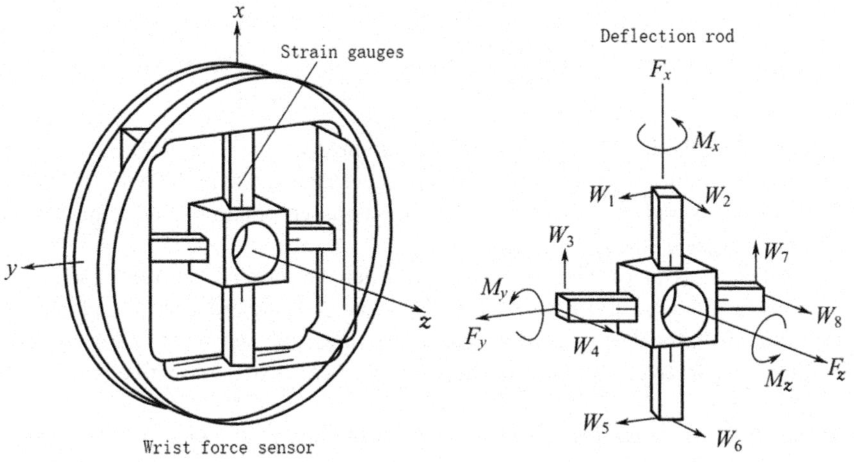

Fig. 3.1 Flexible cross beam wrist force sensor

$$\left.\begin{aligned} F_x &= -f_{x1} - f_{x3} \\ F_y &= -f_{y1} - f_{y2} - f_{y3} - f_{y4} \\ F_z &= -f_{z2} - f_{z4} \\ M_x &= a(f_{z2} + f_{z4}) + b(f_{y1} - f_{y4}) \\ M_y &= -b(f_{x1} - f_{x3} - f_{z2} + f_{z4}) \\ M_z &= -a(f_{x1} + f_{x3} + f_{z2} - f_{z4}) \end{aligned}\right\} \tag{3.1}$$

- Cartridge wrist force sensor

Figure 3.2 shows a cylindrical 6-DOF wrist force sensor, the main body is an aluminum cylinder, and there are 8 beams on the outside to support, of which 4 are horizontal beams and 4 are vertical beams. The strain gauges of the horizontal beam are attached to the upper and lower sides, and the strains of the strain gauges are set as Q_x^+, Q_y^+, Q_x^-, Q_y^-, respectively; while the strain gauges of the vertical beams are attached to the left and right sides, and the strains of the strain gauges are set to be respectively for P_x^+, P_y^+, P_x^-, P_y^-. Then, the 6-dimensional force applied to the sensor, that is, the forces Fx, Fy, Fz in the x, y, and z directions, and the torques Mx, My, and Mz in the x, y, and z directions can be calculated by the following relationship, namely,

$$\left.\begin{aligned} F_x &= K_1(P_y^+ + P_y^-) \\ F_y &= K_2(P_x^+ + P_x^-) \\ F_z &= K_3(Q_x^+ + Q_x^- + Q_y^+ + Q_y^-) \\ M_x &= K_4(Q_y^+ - Q_y^-) \\ M_y &= K_5(-Q_x^+ - Q_x^-) \\ M_z &= K_6(P_x^+ - P_x^- - P_y^+ + P_y^-) \end{aligned}\right\} \tag{3.2}$$

In the formula, Kl, K2, K3, K4, K5, K6 are proportional coefficients, which are related to the strain sensitivity of the strain gauges attached to each beam. The strain amount is measured by a half-bridge circuit composed of strain gauges attached to both sides of each beam.

The characteristic of this structure is that when the sensor is working, each beam is mainly designed with bending strain. Therefore, it has a certain degree of normalization, and reasonable structural design can make the sensitivity of each beam uniform and effectively improve, but the disadvantage is that the structure is relatively complex.

The six-dimensional force sensor is one of the most important external sensors of the robot. The sensor can obtain all the information including 3 forces and 3 moments Therefore, it is widely used in advanced robot control such as force/position control, shaft-hole coordination, contour tracking and dual-robot coordination, and has become an indispensable and important tool to ensure the safety of robot operation and improve the operation ability.

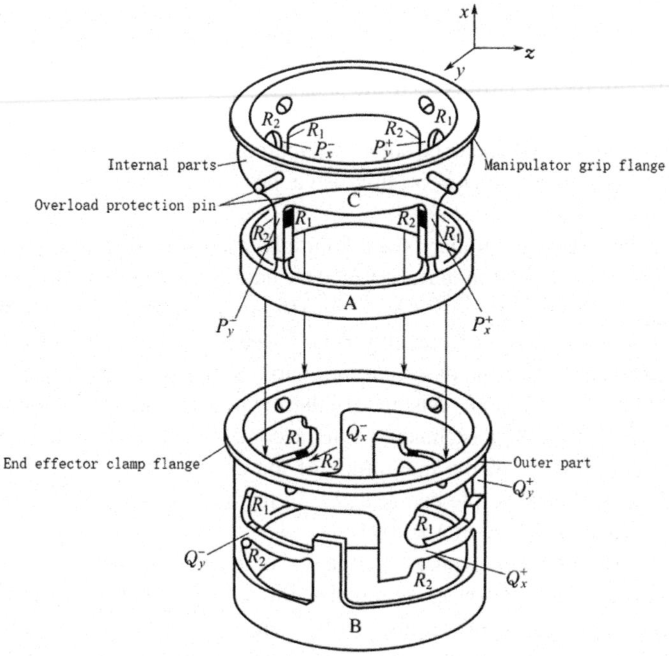

Fig. 3.2 Cartridge 6-DOF wrist force sensor

3.3 Tactile Sensor

Human touch includes touch, pressure, heat and cold, sliding, pain, etc. These perception abilities are very important to human beings and cannot be completely replaced by other perception abilities (such as vision). The robot tactile sensor can realize the functions of touch, pressure and slip, and measure whether there is contact between the gripper and the object being grasped, the contact position and the magnitude of the contact force. A tactile sensor includes a sensor composed of a single sensitive element and a tactile sensor array composed of multiple sensitive elements. When the robot end manipulator is in contact with the external environment, a small displacement can generate a large contact force. This feature is essential for operations that need to eliminate small position errors, such as precision assembly and other occasions that require precise control. Vision is accomplished by the action of light. When the illumination is limited, some simple recognition functions can be accomplished by touch alone. More importantly, touch can also sense the surface features and physical properties of objects, such as softness, hardness, elasticity, roughness, material, etc.

The simplest and earliest used tactile sensor is the microswitch. It has a wide working range, is free from electrical and magnetic interference, is simple, easy to use, and has low cost. A single micro switch usually works in the on and off state, and can indicate whether it is in contact or not in a two-digit manner. If you only need to detect whether it is in contact with the object, this two-digit micro switch can meet the requirements. However, if the shape of the object needs to be detected, it is necessary to install sensitive elements on the contact surface with a high density. Although the micro switch can be small, compared with the requirements of a highly sensitive tactile sensor, such on–off microswitches are still too large for high-density mounting.

Conductive synthetic rubber is a commonly used sensitive element for tactile sensors. It is a conductive material formed by adding conductive particles or semiconductor materials (such as silver or carbon) to silicone rubber. This material is inexpensive, easy to use, and flexible, and can be used on the finger surfaces of robotic multi-fingered dexterous hands. There are various industrial grades of conductive synthetic rubber. The volume resistance of these types of conductive rubbers changes very little when the voltage is changed, but the contact area and reverse contact resistance change greatly with the magnitude of the external force. The tactile sensor fabricated using this principle can realize 256 tactile sensitive units in an area of 1 cm^2, with a sensitive range of 1 to 100 g.

Figure 3.3 shows a piezoresistive tactile sensor using D-section conductive rubber wires, which uses two layers of conductive rubber wires that are perpendicular to each other to achieve row and column cross positioning. When the positive pressure is increased, the D-section conductive rubber is deformed, the contact area increases, and the contact resistance decreases, thereby realizing tactile sensing.

Another commonly used type of tactile sensitive element is the semiconductor strain gauge. Both metallic and semiconductor piezoresistive elements have been used in tactile sensor arrays. Among them, metal foil strain gauges are the most widely used, especially when they are pasted with deformation elements, they can convert external forces into strains, so the strain gauges used for measurement are the most widely used. Using semiconductor technology, strain elements can be fabricated on semiconductors such as silicon, and even signal conditioning circuits can be fabricated on the same silicon wafer. Silicon tactile sensors have the advantages of good linearity, low hysteresis and creep, and the ability to make multiplexing, linearization and temperature compensation circuits in silicon. The disadvantage is that the sensor is prone to overload. In addition, the planar conductivity of silicon

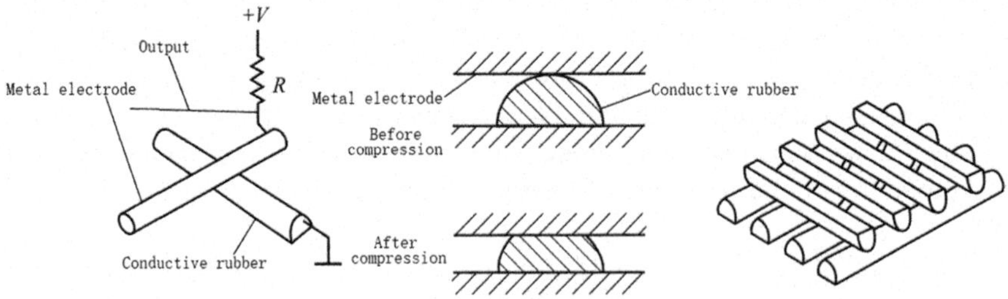

Fig. 3.3 D-section conductive rubber piezoresistive tactile sensor

integrated circuits also limits its application in robotic dexterous fingertip shape sensors.

Some crystals have piezoelectric effect and can also be used as a type of tactile sensitive element, but crystals are generally brittle, and it is difficult to directly make tactile or other sensors. Polymers such as PVF2 (polyvinylidene fluoride) discovered in 1969 have good piezoelectricity, especially flexibility, so they are ideal tactile sensor materials. Certainly, there are many methods and basis for making robot tactile sensors, such as optical, magnetic, capacitive, ultrasonic, chemical and other principles, it is possible to develop robot tactile sensors.

- Piezoelectric sensor

A commonly used piezoelectric crystal is a quartz crystal, which generates a certain electrical signal when subjected to pressure. The strength of the electrical signal output by the quartz crystal is determined by the pressure value it receives. By detecting the strength of these electrical signals, the force on the object to be measured can be detected. Piezoelectric force sensors can not only measure the pressure on the object, but also the tensile force. When measuring the tensile force, it is necessary to give a certain preload force to the piezoelectric crystal. Since piezoelectric crystals cannot withstand excessive strain, they have a small measurement range. In robotic applications, there is generally no excessive force, so piezoelectric force sensors are more suitable. When the piezoelectric sensor is installed, the parts in contact with the sensor surface should have good parallelism and low surface roughness, and its hardness should also be lower than the hardness of the sensor contact surface to ensure that the preload is perpendicular to the sensor surface, so that a uniform distribution of pressure is created on the quartz crystal. Figure 3.4 shows a three-component force piezoelectric sensor. It consists of three pairs of quartz wafers, capable of measuring forces in three directions simultaneously. The upper and lower pairs of wafers use the shearing effect of the crystal to measure the force in the x-direction and the y-direction respectively; the middle pair of wafers use the longitudinal piezoelectric effect of the crystal to measure the force in the z-direction.

- Optical fiber pressure sensor

The optical fiber pressure sensor unit shown in Fig. 3.5 is based on the principle of total internal reflection destruction, and is a high-sensitivity optical fiber sensor that realizes light intensity modulation. The sending fiber and the receiving fiber are connected by a right-angle prism, and the air gap between the prism slope and the displacement diaphragm is about 0.3 μm. The lower surface of the diaphragm is coated with a light absorbing layer. When the diaphragm moves downward under pressure, the air gap between the prism slope and the light absorbing layer changes, which causes local destruction of the total (internal) reflection in the prism interface, causing some light leaves the upper interface into the absorbing layer and is absorbed, so the light intensity in the receiving fiber changes accordingly. The light absorbing layer can be made of glass material or silicone rubber with good plasticity, and is made by coating method.

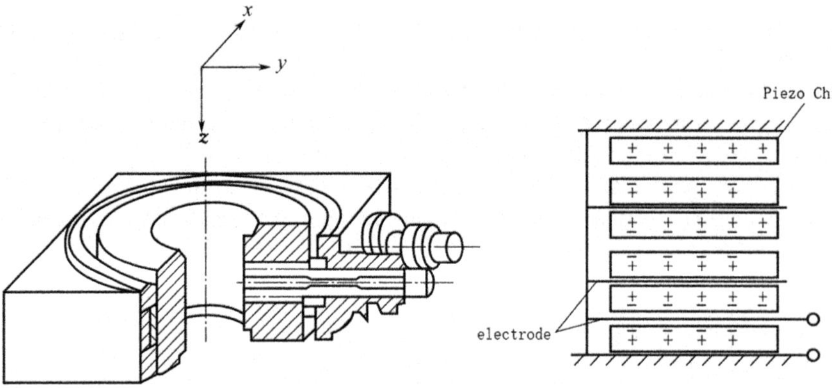

Fig. 3.4 Three-component force piezoelectric sensor

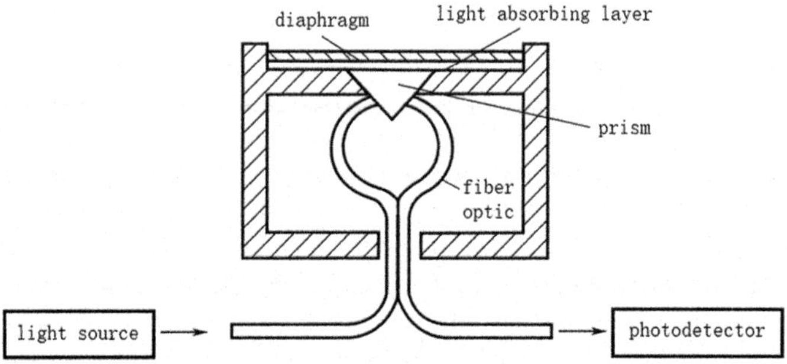

Fig. 3.5 Fiber pressure sensor unit

When the diaphragm is compressed, it will bend and deform. For the diaphragm fixed at the periphery, when the deflection is small (W ≤ 0.5t), the deflection of the center of the diaphragm is calculated as follows:

$$W = \frac{3(1 - \mu^2)a^4 p}{16Et^3} \tag{3.3}$$

In the formula, W is the central deflection of the diaphragm; E is the elastic modulus; t is the thickness of the diaphragm; μ is the Poisson's ratio; p is the pressure; a is the effective radius of the diaphragm.

Equation (3.3) shows that under the condition of small load, the central displacement of the diaphragm is proportional to the pressure.

- Slip sensor

When the robot grasps objects with unknown properties, it should be able to determine the given value of the optimal grip force by itself. When the gripping force is not enough, it is necessary to detect the sliding of the object being grasped. Using the

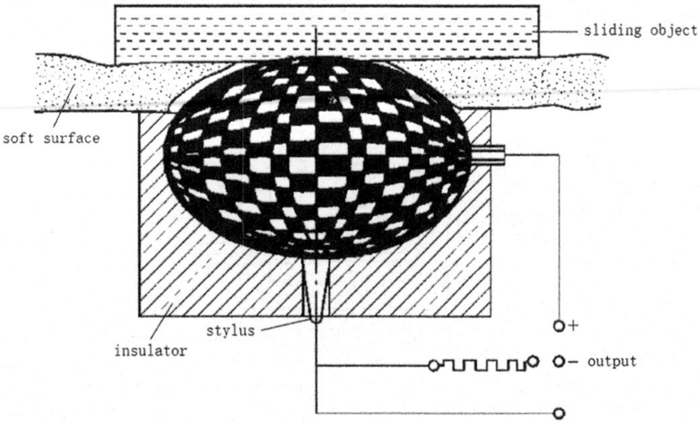

Fig. 3.6 Ball type slide sensor

detection signal, consider the most reliable clamping method without damaging the object. The sensor that realizes this function is called a sliding sensor.

There are rolling and ball types of slip sensors, as well as a sensor that detects slip through vibration. When an object slides on the sensor surface, it comes into contact with the rollers or rings, turning the sliding into a rotation. Figure 3.6 shows a ball-type slip sensor developed by the University of Belgrade, Yugoslavia, consisting of a metal ball and a stylus. The surface of the metal ball is divided into a plurality of conductive and insulating grids arranged in phases, and the tip of the contact pin is small and can only touch one grid at a time. When the workpiece slides, the metal ball also rotates, and a pulse signal is output on the stylus. The frequency of the pulse signal reflects the sliding speed, and the number of pulse signals corresponds to the sliding distance.

Figure 3.7 shows the vibrating slip sensor, the steel ball pointer sticks out of the sensor and contacts the object. When the workpiece moves, the pointer vibrates and the coil outputs a signal. The use of rubber and oil as a damper reduces the sensitivity of the sensor to vibrations of the robot itself.

3.4 Vision Sensor

In order to make the service robot have the function of autonomous action, the service robot should have the ability to recognize the outside world, especially the operation object must be recognized. Among the information obtained by the service robot from the outside world, the biggest information is visual information, and the visual sensor is a remote sensor that can detect without contacting the object. Although two-dimensional image processing is performed on the outside world, 3D information can also be recognized if appropriate information processing is performed.

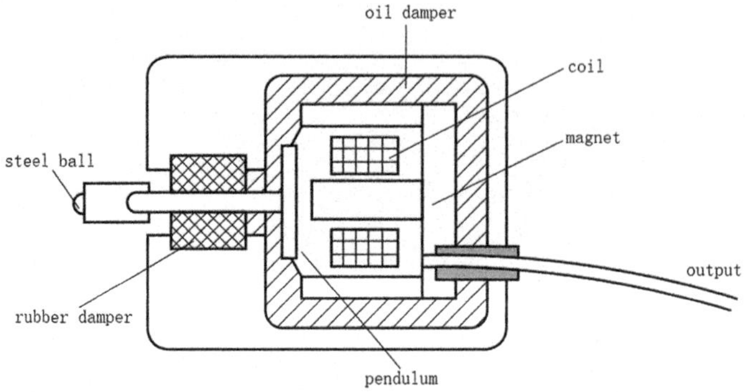

Fig. 3.7 Vibrating slide sensor

- Photoelectric conversion device

In the artificial vision system, the photoelectric conversion devices equivalent to the visual cells of the eye include photodiodes, phototransistors, and CCD image sensors. The tube-shaped photoelectric conversion devices used in the past have been gradually replaced by solid-state devices with the development of semiconductor technology due to their high operating voltage, high power consumption and large size.

– Photo Diode

When the semiconductor PN junction is irradiated by light, if the photon energy is greater than the forbidden band width of the semiconductor material, the photon will be absorbed to form an electron-hole pair, a potential difference will be generated, and a current or voltage corresponding to the amount of incident light will be output. A photodiode is a light sensor that utilizes the photovoltaic effect, and Fig. 3.8 shows its volt-ampere characteristics. When a photodiode is used, a reverse bias voltage is generally applied, and it can be used without a bias voltage. At zero bias, the PN junction capacitance becomes larger and the frequency response decreases, but the linearity is good. If reverse bias is applied, the depletion layer without carriers increases, and the response characteristics improve. Depending on the circuit structure, the response time of light detection can be below 1 ns.

In order to improve the resolution of distance measurement with lidar, photoelectric conversion elements with good response characteristics are required. The avalanche photodiode (APD) is developed by utilizing the amplification principle of electron avalanche generated by electron avalanche caused by the acceleration of the electron's motion under the action of a strong electric field. It is a light sensor that detects weak light and has good response characteristics. As a position detection element, the photodiode can continuously detect the incident position of the light beam, and can also be used for the position detection of the light spot on a two-dimensional plane. Its electrodes are not conductors, but uniform resistive films.

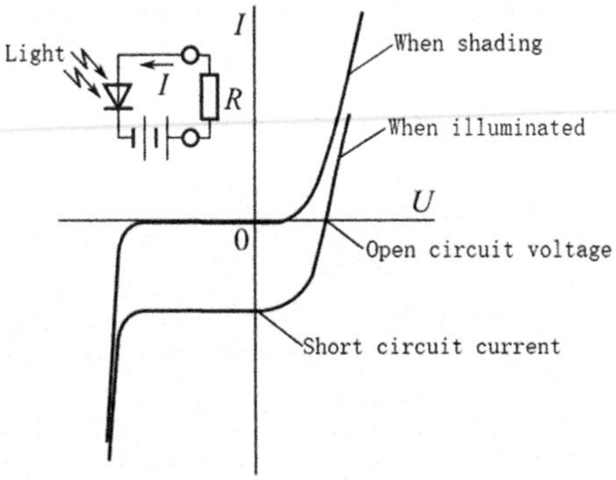

Fig. 3.8 Voltammetric characteristics of photodiode

– Photo Transistor

A photodiode is formed between the collector C and the base B of the PNP or NPN phototransistor. When illuminated by light, a current is generated between the reverse biased base and collector, and the amplified current flows through the collector and emitter. Because the phototransistor has an amplifying function, the photocurrent generated is 100–1000 times that of the photodiode, and the response time is on the order of μs.

– CCD image sensor

CCD is the abbreviation of Charge Coupled Device, which is an element that stores and transmits charges through potential wells. The CCD image sensor adopts the MOS structure and has no PN junction inside. As shown in Fig. 3.9, there is a layer of SiO2 insulating layer on the P-type silicon substrate, and a plurality of metal electrodes are arranged on it. When a positive voltage is applied to the electrode, a potential well is generated under the electrode, and the depth of the potential well changes with the voltage. If the voltage applied to the electrode is changed in turn, the potential well moves with the change of the voltage, and the charge injected into the potential well is transferred. According to the configuration of the electrodes and the change of the driving voltage phase, there are two-phase clock driving and three-phase clock driving transmission methods.

The CCD image sensor is equipped with photosensitive elements and charge transfer devices on a silicon substrate. Through the sequential transfer of charges, the information of a plurality of pixels is time-divisionally and sequentially extracted. This sensor has a one-dimensional line image sensor and a two-dimensional area image sensor. The two-dimensional area image sensor needs to scan in two directions, horizontal and vertical. There are frame transfer methods and inter-line transfer methods. The principle is shown in Fig. 3.10.

– MOS image sensor

Photodiodes and MOS field effect transistors are arranged in pairs on a silicon substrate to form a MOS image sensor. The position of the pixel is determined by

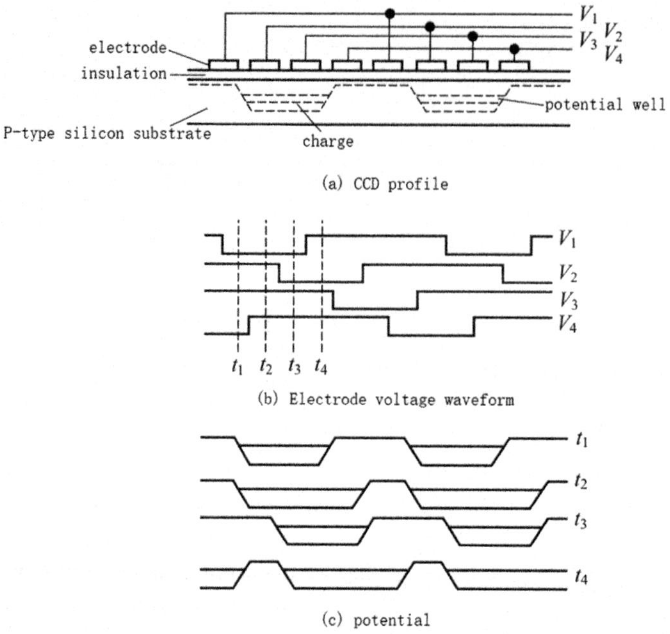

Fig. 3.9 CCD image sensor

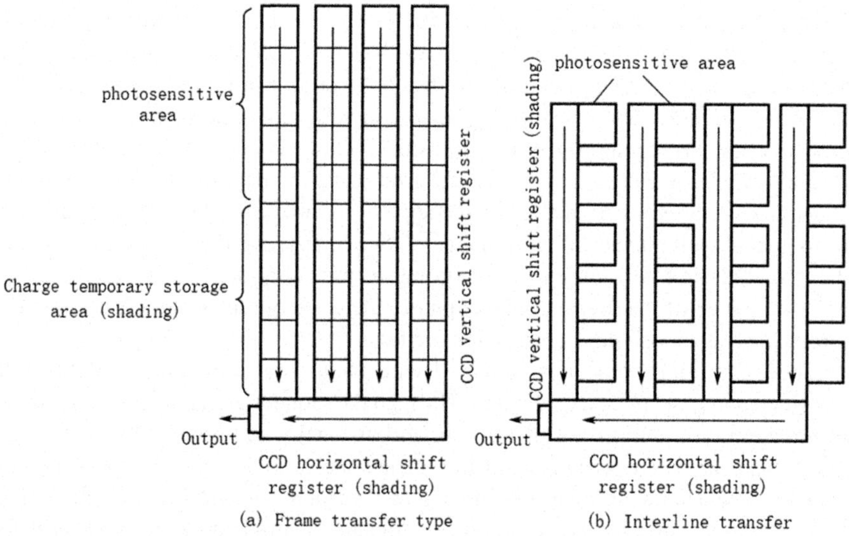

Fig. 3.10 Signal scanning principle of CCD image sensor

selecting the horizontal scanning line and the vertical scanning line, the FET (field-effect tube) on the intersection of the two scanning lines is turned on, and then the pixel information is taken out from the paired photodiode. Scanning is time-sharing and sequential.

– Industrial TV Camera

Industrial TV cameras are composed of two-dimensional surface image sensors and peripheral circuits such as scanning circuits. As long as the power supply is connected, the camera can output the standard TV signal of the captured image. Most camera lenses can be replaced with a 1/2-inch thread called a C-lens mount. In order to realize the automatic focusing of the lens, most photographic lenses have a drive terminal for the automatic iris. Some cameras on the market now have external synchronization signal input terminals for controlling vertical scanning or horizontal and vertical scanning; some can change the charge accumulation time of the CCD to shorten the exposure time. Most of the color cameras are single-panel cameras in which red (R), green (G), and blue (B) color filters are embedded on the image sensor to extract color signals. When the light source is different and the color needs to be adjusted, the method is very simple, just switch manually.

- Two-dimensional vision sensor

Vision sensors are divided into two categories: two-dimensional vision and three-dimensional vision sensors. A two-dimensional vision sensor is a sensor that acquires scene graphics information. The processing methods include binary image processing, grayscale image processing and color image processing, all of which take the input two-dimensional image as the recognition object. The image is acquired by the camera. If the object passes a fixed position at a certain speed on the conveyor belt, the input signal of the two-dimensional image can also be acquired by a one-dimensional line sensor.

For a production line with limited operating objects and adjustable working environment, an inexpensive binary image vision system with short processing time is generally used. In image processing, first of all, it is necessary to distinguish the image as the object image and the bottom as the background image. The distinction between the figure and the bottom is still easy to handle. In figure recognition, data such as the area, perimeter, and center position of the figure should be used. In order to reduce the workload of image processing, the following points must be noted.

– Lighting direction

Not only are there lighting sources in the environment, but there are other lights as well. Therefore, in order to make the brightness of the object and the change of the illumination direction as small as possible, it is necessary to pay attention to the reflected light on the surface of the object and the shadow of the object.

– Contrast of the background

The black object is placed on a white background, and the contrast between the picture and the bottom is large, and it is easy to distinguish. Sometimes a light source is placed behind the object, and the light passes through the diffuser to hit the object to obtain a contour image.

– Location of vision sensor

Change the distance between the vision sensor and the object, and the image size changes accordingly. Changing the viewing direction when acquiring a stereoscopic image changes the shape of the image. Observe the object in the vertical direction to obtain a stable image.

– Placement of objects

If the objects are placed overlapping, it is more difficult to perform image processing. By placing each object separately, image processing time can be shortened.

- 3D Vision Sensor

The three-dimensional vision sensor can obtain the stereoscopic or spatial information of the scene. The stereo image can be obtained according to the data of the inclination direction of the surface of the object and the height distribution of the unevenness, or it can be obtained according to the distribution of the distance from the observation point to the object, that is, the distance image. Spatial information is obtained from distance images. It can be divided into the following categories.

– Monocular observation method

People can understand the depth of field and the concave-convex state of objects by looking at a photo. It can be seen that the state of the surface of the object (texture analysis), the distribution of reflected light intensity, the contour shape, the shadow, etc. are all clues to the stereoscopic information existing in an image. Therefore, one of the current research topics is how to use the knowledge base for image processing to use a TV camera as a stereo vision sensor according to a series of assumptions.

– Moiré fringe method

The Moiré fringe method uses striped light to shine on the surface of an object, and then captures an image through a light-shielding stripe of the same shape at another location. The fringe image and the shading image on the object are offset to form a contour pattern, that is, Moiré fringes. According to the shape of the moiré fringes, the information of the surface concavity and convexity of the object is obtained. Distance can be measured based on the number of stripes, but it is sometimes difficult to determine the number of stripes.

– Active Stereo Vision Method

The light beam shines on the surface of the target object, and the image of the object is captured at a certain distance from the baseline, and the position of the light spot is detected from it, and then the distance of the light spot is obtained according to the principle of triangulation. This method of obtaining stereo information is the active stereo vision method.

– Passive Stereoscopic Vision

Passive stereo vision method is like two human eyes, finds the position of the image of the same object point from two images obtained from different sight lines, and uses the principle of triangulation to obtain the distance image. Although this method is simple in principle, it is very difficult to detect the corresponding point of the same object point in two images. Passive vision uses natural measurements, such as binocular vision, which is passive vision.

– Lidar

Use laser instead of radar wave, scan in the field of view, and obtain distance image by measuring the return time of reflected light. It can be divided into two methods: one emits a pulsed beam, receives the reflected light with a photomultiplier tube, and directly measures the return time of the light; the other emits an amplitude-modulated laser and measures the phase lag of the modulated waveform of the reflected light. To improve the distance resolution, the temporal resolution of reflected light detection must be improved, thus requiring cutting-edge electronics.

3.5 Auditory Sensor

• The sound mechanism of speech

The pitch, timbre and loudness of speech constitute the three elements of speech. The tone is mainly related to the frequency of the sound wave, and they have a logarithmic relationship. The timbre is related to the spectral structure of the sound wave and the analog waveform. The loudness is positively related to the amplitude of the sound wave signal.

The physical essence of speech is sound waves, which are longitudinal waves and vibration waves. After the sound source vibrates, the propagation medium around the sound source vibrates physically, and the sound wave spreads with the vibration of the propagation medium.

Sound is a wave phenomenon produced by the vibration of an object, transmitted through a gas, solid or liquid, and sensed by the sound-sensing organ. Human speech is produced by the physiological movement of human vocal organs under the control of the brain. The human vocal organ consists of three parts:

– Lungs and trachea produce sound source: The lungs produce compressed air, which is sent to the sound generation system through the trachea; the trachea connects the lungs and the larynx, and is the channel connecting the lungs with the vocal tract.

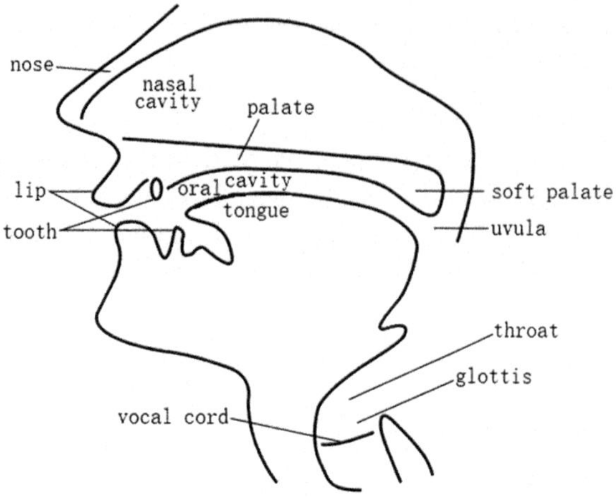

Fig. 3.11 Schematic diagram of sound channel profile

- The larynx and the vocal cords form the glottis: the larynx is a complex system that controls the movement of the vocal cords; the acoustic function of the vocal cords is mainly to generate excitation.
- The pharyngeal cavity, oral cavity and nasal cavity form the vocal tract: the oral cavity and the nasal cavity are the resonators during sound production. The organs in the oral cavity act together to form different obstacles when the air flow passes through, which in turn generates different vibrations and produces different sounds.

Air is expelled from the lungs into the larynx, through the vocal cords into the vocal tract, and finally radiates sound waves from the nose, ultimately forming speech. Figure 3.11 shows a schematic diagram of the channel section.

- Auditory sensors and language recognition

An auditory sensor, that is, a sound-sensitive sensor, is a device or equipment that converts the mechanical vibration of sound waves into electrical signals. There are many kinds of sound sensors, which can be divided into piezoelectric, electrostrictive effect, electromagnetic effect, electrostatic effect and magnetostriction according to the measurement principle. A common auditory sensor is a condenser electret microphone, which is a piezoelectric sensor. The sensor has a built-in condenser electret microphone that is sensitive to sound. The sound wave vibrates the electret film in the sensor, resulting in a change in capacitance and a small voltage corresponding to the change. The voltage signal undergoes analog-to-digital conversion and is converted into a digital signal that can be recognized by the computer. Ultimately, the auditory sensor converts the mechanical vibration signal of the sound into a digital signal that the computer can recognize and calculate.

The auditory sensor obtains the pitch, timbre and loudness information of speech by perceiving the vibration information of sound waves. The perception of speech is mainly to perceive the pitch and loudness information of speech. To ensure that the service robot can work safely, it is often necessary to install hearing sensors. Because the vision sensor can't monitor the full range of 360°, but the auditory sensor can carry out full range monitoring. It is more convenient for humans to command industrial robots with language than with keyboards. Therefore, auditory sensors are needed to detect various sounds made by humans, and then the language recognition system can identify and execute commands.

– Hearing sensor

The function of the hearing sensor is to convert the acoustic signal into an electrical signal, which is also called a microphone. Commonly used hearing sensors include moving coil sensors and capacitive sensors.

a. Moving coil microphone. Figure 3.12 shows the structure principle of the moving coil microphone. The diaphragm of a microphone is very light, thin, and vibrates with the sound. The moving coil is glued to the diaphragm and can move with the vibration of the diaphragm. The moving coil floats in the magnetic field of the magnetic gap. When the moving coil moves in the magnetic field, an induced electromotive force can be generated in the moving coil. This electromotive force corresponds to the diaphragm and frequency, so the electrical signal output by the moving coil corresponds to the strength of the sound and the level of the frequency. Through this process, the microphone converts the sound into an audio electrical signal for output.
b. Condenser microphone. Figure 3.13 shows the structure principle of a condenser microphone. A capacitor is formed by a fixed electrode and a diaphragm, and a polarization voltage is applied to the fixed electrode of the capacitor through a resistor R_L. When sound is introduced, the diaphragm can vibrate with time, and the capacitance between the diaphragm and the fixed electrode also changes with the sound, and the impedance of this capacitor also changes; the resistance of the load resistor R_L in series with it is fixed, the impedance change of the capacitor is shown as the change of the potential at point a. Through the coupling capacitor

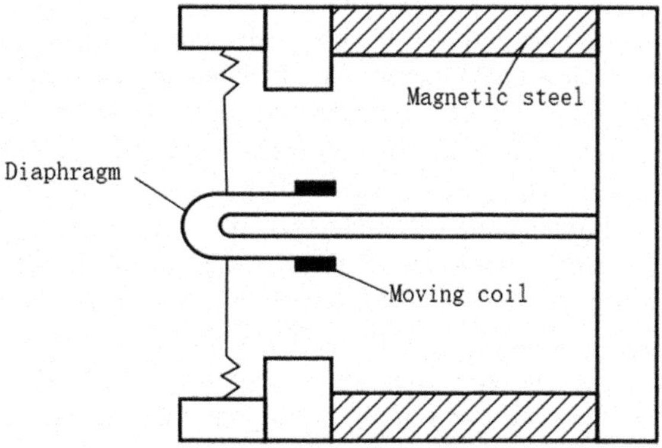

Fig. 3.12 Structural principle of moving coil microphone

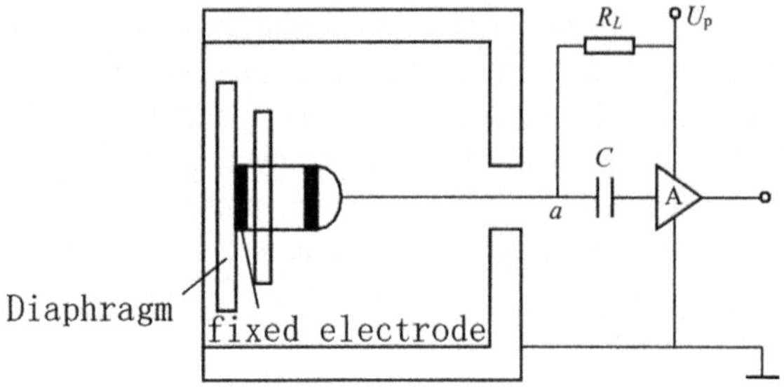

Fig. 3.13 The structure principle of condenser microphone

C, the signal of the resistance change at point a is input to the preamplifier A, and the audio signal is output after being amplified.

- Speech recognition chip. Speech recognition technology is an advanced technology that allows robots to convert the speech signals collected by sensors into corresponding text or commands through the process of recognition and understanding. The computer speech recognition process is basically the same as the human speech recognition process. At present, the mainstream speech recognition technology is based on the basic theory of statistical pattern recognition, and a complete speech recognition system can be roughly divided into three parts.

a. Voice feature extraction. Its purpose is to extract time-varying speech feature sequences from speech waveforms. The extraction and selection of acoustic features is an important part of speech recognition. The extraction of acoustic features is a process of greatly compressing information in order to make the mode divider better. Due to the time-varying nature of speech signals, feature extraction must be performed on a small segment of speech signals, that is, short-term analysis.
b. Identification algorithm. The acoustic model is the underlying model of the recognition system and is the most critical part of the speech recognition system. The acoustic model is usually generated by training the acquired speech features, and the purpose is to establish a pronunciation template for each pronunciation. During recognition, the unknown speech features are matched and compared with the acoustic model (pattern), and the distance between the feature vector sequence of the unknown speech and each pronunciation template is calculated. The design of acoustic models is closely related to the characteristics of language pronunciation. The unit size of the acoustic model (word pronunciation model, semi-syllabic model or phoneme model) has a great influence on the size of speech training data, system recognition rate and flexibility.
c. Semantic understanding. The computer performs grammatical and semantic analysis on the recognition results, understands the meaning of the language, and responds accordingly, usually through a language model.

- Microphone array

Usually people see the microphone can identify the strength of the sound, it is also called the sound/hearing sensor. A microphone array is a group of omnidirectional microphones located at different positions in space, which are arranged according to certain rules to form an array, which is a kind of auditory sensor for spatial acquisition of sound signals propagated in space. The voice information collected by the microphone array includes not only the pitch and loudness information of the voice, but also the spatial position information of the voice. According to the distance between the sound source and the microphone array, the microphone array can be divided into near-field model and far-field model; according to the topology of the microphone array, the microphone array can be divided into linear array, plane array and volume array.

– Near-field model and far-field model

There is no absolute standard for the division of the near-field model and the far-field model. Generally, it is considered that the distance between the sound source and the reference point of the center of the microphone array is greater than the signal wavelength, which is the far-field, on the contrary, the near-field.

As shown in Fig. 3.14, if the distance from the sound source to the center of the microphone array is greater than r, it is a far-field model, otherwise it is a near-field model. S is the physical center point of the microphone array, that is, the reference point of the microphone array.

The distance from the sound source to the center of the microphone array:

$$\mathrm{r} = 2d^2/\lambda_{\min} \tag{3.4}$$

in the formula, d—Distance between adjacent microphones in a uniform linear array, m;

$\lambda_{\min}$—The wavelength of the highest frequency speech of the sound source, m.

The near-field model regards the sound wave as a spherical wave and considers the amplitude difference between the signals received by the microphone elements; the far-field model regards the sound wave as a plane wave, ignoring the amplitude difference between the signals received by each array element. The far-field model

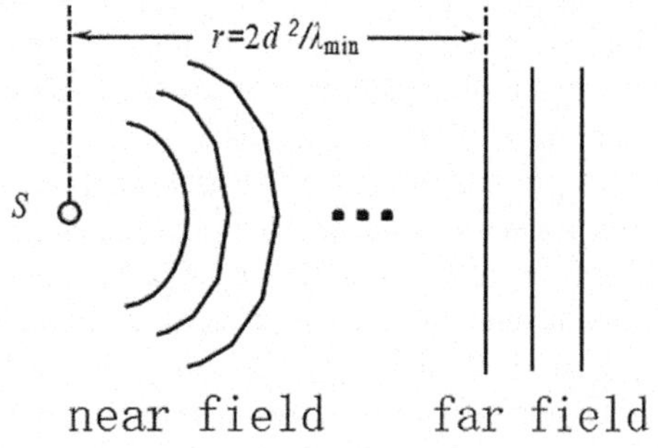

Fig. 3.14 Near-field and far-field models

is a simplification of the actual model, and it is considered that there is a delay relationship between the received signals.

– Microphone array topology

According to the dimension of the microphone array, it can be divided into one-dimensional, two-dimensional and three-dimensional microphone arrays.

A one-dimensional microphone array, that is, a linear microphone array, has the centers of its array elements located on the same straight line. According to whether the distance between adjacent array elements is the same, it can be divided into uniform linear array and nested linear array. The linear microphone array can only obtain the horizontal direction angle information of the spatial position of the speech signal.

A two-dimensional microphone array, that is, a plane microphone array, has the center of the array elements distributed on a plane. According to the geometry of the array, it can be divided into equilateral triangle array, T-shaped array, uniform circular array, uniform square array, coaxial circular array, rectangular area array, etc. The two-dimensional microphone array can obtain the horizontal azimuth and vertical azimuth information of the spatial position information of the speech signal.

A three-dimensional microphone array, that is, a stereo microphone array, has the centers of its array elements distributed in a stereo space. According to the spatial shape of the microphone array, the three-dimensional microphone array can be divided into a tetrahedral array, a square array, a rectangular array, a spherical array, and the like. The three-dimensional microphone array can obtain the horizontal azimuth angle, vertical azimuth angle, and distance information of the sound source and the reference point of the microphone array of the spatial position of the speech signal.

3.6 Olfactory Sensor

Olfactory sensors mainly use gas sensors, ray sensors, etc., which are mostly used to detect chemical components and concentrations in the air. In the harsh environment of radiation, high temperature gas, combustible gas and other toxic gases, it is very important to develop sensors for detecting radiation, combustible gases and toxic gases. It is of great significance to people's understanding of environmental pollution, prevention of fire and gas leakage alarm.

A gas sensor is a device that detects a specific component in a gas (mostly air) and converts it into an electrical signal to provide information about the presence and concentration of the gas to be measured.

Gas sensors were first used for flammable gas leakage alarms, disaster prevention and production safety; later, they were gradually popularized and applied for the detection of toxic gases, leak detection of containers or pipes, environmental monitoring (to prevent pollution), combustion detection and control of boilers and automobiles (can save fuel and reduce harmful gas emissions), detection and automatic control of industrial processes (measure and analyze the content or concentration of a certain gas in the production process). In recent years, gas sensors have been widely used in medical treatment, air purification, household gas stoves and water heaters.

The performance of the gas sensor must meet the following conditions:

- It can detect the allowable concentration of explosive gases, the allowable concentration of harmful gases and other reference set concentrations, and can give alarm, display and control signals in time.
- It is not sensitive to coexisting gases or substances other than the measured gas.
- Long-term stability of performance. Good repeatability.
- Good dynamic characteristics and quick response.
- It is easy to use and maintain, and the price is cheap.

- Surface-controlled gas sensor

The change in the surface resistance of such devices depends on the exchange of electrons between the originally adsorbed gas on the surface and the semiconductor material. Usually the device works in the air, the oxygen in the air and the gas with high electron compatibility such as NO_2 accept the electrons from the semiconductor material and absorb the negative charge. The result is that the conduction electrons in the surface space charge area of the N-type semiconductor material are reduced, so that the surface conductivity is reduced, thereby placing the device in a high resistance state. Once the device is in contact with the measured gas, it will react with the adsorbed oxygen and release several electrons bound by oxygen, which increases the surface conductance of the sensitive film and reduces the device resistance. Most of this type of sensors use flammable gas as the detection object, but if the adsorption capacity is strong, even non-flammable gas can be used as the detection object. It has high detection sensitivity. It has the advantages of fast response speed and great practical value.

- Contact combustion gas sensor

Generally, the gas that reaches a certain concentration in the air and can cause combustion when it touches the fire is called flammable gas. Such as methane, acetylene, methanol, ethanol, ether, carbon monoxide and hydrogen are all flammable gases.

The contact combustion type gas sensor consists of a metal coil such as platinum embedded in an oxidation catalyst. When using, apply current to the metal coil to keep it in a high temperature state of 300–600 °C, the component is connected to a bridge arm in the bridge circuit at the same time, and the bridge circuit is adjusted to make it balanced. Once the combustible gas comes into contact with the surface of the sensor, the heat of combustion further increases the temperature of the metal wire, causing the resistance of the device to increase, thereby destroying the balance of the bridge. The unbalanced current or voltage it outputs is proportional to the concentration of combustible gas, and the concentration of combustible gas can be measured by detecting this current and voltage.

The advantages of the contact combustion gas sensor are good gas selectivity, good linearity, little influence by temperature and humidity, and fast response. The disadvantage is that the sensitivity to low-concentration combustible gas is low, the characteristics of the sensitive element are sharply reduced after being attacked by the catalyst, and the metal wire is easily broken.

- Smoke sensor

Smoke is formed by the suspension of particles much larger than gas molecules in the gas. Different from the analysis of general gas components, it is necessary to use the characteristics of particles to detect. This type of sensor is mostly used in fire alarms, and it is also a sensor that determines the output signal based on the presence or absence of smoke, and cannot be continuously measured quantitatively.

– Scattering

A shading screen is set between the light-emitting tube and the photosensitive element. The photosensitive element cannot receive light signals without smoke. When there is smoke, the photosensitive element can send out electrical signals by means of the scattered light of the particles. The principle is shown in Fig. 3.15. The sensitivity of this sensor is independent of the type of smoke.

– Ionic

The radioactive isotope americium Am241 is used to emit a small amount of alpha rays to ionize the nearby air. When there is a DC voltage between the parallel plate electrodes, an ionic current is generated. When there is smoke, the particles will adsorb the ions, and the ions themselves also absorb alpha rays, resulting in a decrease in the ion current.

If there is an ion chamber sealed with pure air as a reference element, the ion currents of the two can be compared to eliminate external interference and obtain

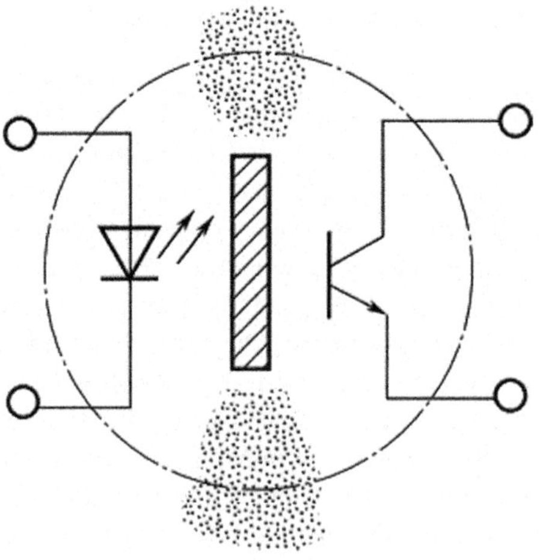

Fig. 3.15 Working principle of scattering smoke sensor

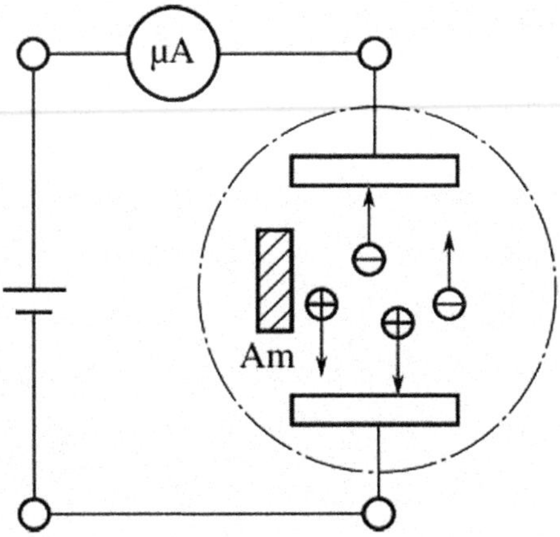

Fig. 3.16 The working principle of the ionization smoke sensor

reliable detection results. The sensitivity of this method is related to the type of smoke. The working principle can be seen in Fig. 3.16.

3.7 Proximity Sensor

Proximity sensors are sensors used by robots to detect the relative position and distance between themselves and surrounding objects. Its use is of great significance to timely trajectory planning and accident prevention during robot work. It mainly plays the role of the following three aspects.

- Get the necessary information before touching the object and prepare for the next action.
- When an obstacle is found, change the path or stop to avoid collision.
- Obtain information about the surface shape of the object.

According to the sensing range (or distance), proximity sensors can be roughly divided into three categories: magnetic (inductive) for sensing close-range objects (mm-level), air pressure, capacitive, etc.; infrared photoelectric type to perceive objects at medium distance (within about 30 cm); ultrasonic type and laser type for sensing objects at a distant (30 cm away). Vision sensors can also act as proximity sensors.

- Magnetic proximity sensor

Figure 3.17 shows the structure principle of the magnetic sensor. It is composed of excitation coil C_0 and detection coils C_1 and C_2. C_1 and C_2 have the same number of turns and are connected in a differential type. When it is not close to the object, the output is 0 due to the symmetry of the structure. When it is close to the object (metal), the magnetic flux changes due to the eddy current generated by the metal,

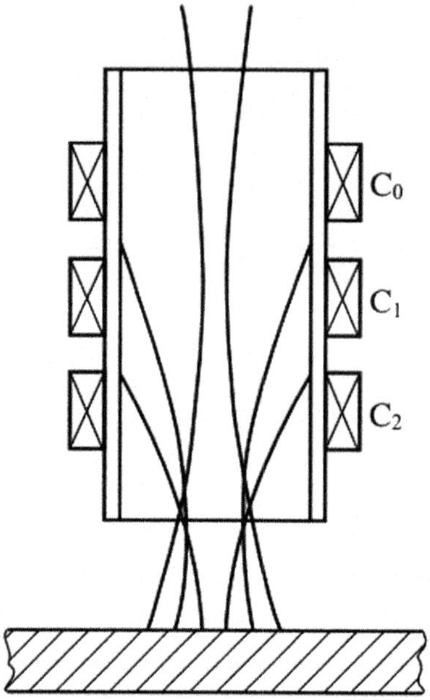

Fig. 3.17 Magnetic proximity sensor structure principle

so that the output of the detection coil changes. This sensor is not affected by light, heat, and surface features of objects, and can be miniaturized and lightweight, but it can only detect metal objects.

Hitachi of Japan uses it on arc welding robots to track welds. The detection distance is 0–8 mm below 200 °C, and the error is only 4%.

- Pneumatic proximity sensor

Figure 3.18 is the basic principle and characteristic diagram of the air pressure sensor. It is designed according to the principle of nozzle-baffle action. The air pressure source p_V enters the back pressure chamber through the orifice, and then is ejected through the nozzle. After the airflow hits the measured object, the back pressure output p_A is formed. Reasonable selection of p_V value (constant pressure source), nozzle size and orifice size, the corresponding relationship between output p_A and distance x can be obtained. Generally, it is not linear, but local approximate linear output can be achieved. This kind of sensor has strong fire prevention, anti-magnetic, and anti-radiation capabilities, but it requires the air source to maintain a certain degree of purification.

- Infrared proximity sensor

The infrared sensor is a relatively effective proximity sensor. The wavelength of the light emitted by the sensor is about a few hundred nanometers, which is a short-wavelength electromagnetic wave. It is a radiant energy converter, which is mainly used to convert the received infrared radiant energy into other forms of energy such as electrical energy, thermal energy, etc., which are easy to measure or observe.

Fig. 3.18 The basic principle and characteristics of pneumatic proximity sensor

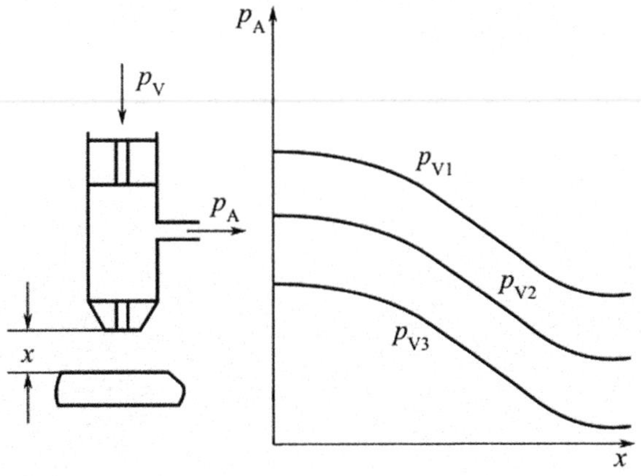

According to the energy conversion method, infrared detectors can be divided into two categories: thermal detectors and photon detectors. Infrared sensors are not interfered by electromagnetic waves, non-noise sources, and can achieve very contact measurements. In addition, infrared rays (referring to middle and far infrared rays) are not affected by surrounding visible light, so measurements can be performed day and night.

Similar to the sonar sensor, the infrared sensor works in the transmit/receive state. This sensor emits infrared light from the same emission source and uses two photodetectors to measure the amount of light reflected back. Because these instruments measure light differently, they are greatly affected by the environment, the color of the object, the direction, and the surrounding light can all contribute to measurement errors. However, since the emitted light is light rather than sound, more infrared sensor measurements can be obtained in a relatively short period of time, and the distance measurement range is relatively short.

The infrared sensor ranging based on the principle of triangulation is now introduced. That is, the infrared transmitter emits an infrared beam at a certain angle. When it encounters an object, the beam will be reflected back, as shown in Fig. 3.19. After the reflected infrared light is detected by the CCD detector, an offset value L will be obtained. Using the triangular relationship, after knowing the emission angle α, the offset distance L, the center moment X, and the focal length f of the filter, The distance D from the sensor to the object can be calculated through the geometric relationship.

It can be seen that when the distance of D is close enough, the value of L will be quite large, which exceeds the detection range of the CCD. At this time, although the object is very close, the sensor cannot see it. When the object distance D is large, the L value will be small. At this time, whether the CCD detector can distinguish this small L value becomes the key, that is to say, the resolution of the CCD determines whether a sufficiently accurate L value can be obtained. To detect the farther the object, the higher the resolution requirements of the CCD.

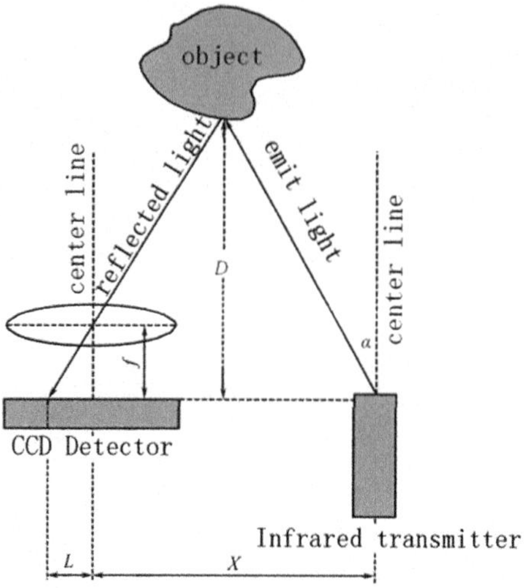

Fig. 3.19 Schematic diagram of infrared sensor ranging

The output of this sensor is non-linear. As can be seen from Fig. 3.20, when the distance of the detected object is less than 10 cm, the output voltage drops sharply, that is to say, from the voltage reading, the distance of the object should be getting farther and farther. But this is not the case. If the robot is slowly approaching the obstacle and suddenly cannot detect the obstacle, in general, the control program will make the robot move at full speed, and the result is that the robot hits the obstacle. The solution to this problem is to change the installation position of the sensor so that the distance from the robot's periphery is greater than the minimum detection distance, as shown in Fig. 3.21.

Affected by the characteristics of the device, the infrared sensor has poor anti-interference, that is, it is easily affected by various heat sources and ambient light. The color and surface smoothness of the detected object are different, and the intensity of the reflected infrared rays will be different. And due to the influence of the sensor power factor, its detection distance is generally between 10 and 500 cm.

- Ultrasonic distance sensor

Ultrasonic proximity sensors are used for robots to detect the presence and distance of surrounding objects. Especially for mobile robots, the installation of this sensor can detect whether there are obstacles on the way forward at any time to avoid collisions.

Ultrasound is a kind of mechanical wave that cannot be heard by human ears. Its frequency is above 20 kHz, its wavelength is short, and its diffraction is small, so it can be directionally propagated as a ray. Ultrasonic sensor consists of ultrasonic generator and receiver. Ultrasonic generators include piezoelectric, electromagnetic and hysteresis telescopic. The most commonly used detection technology is piezoelectric. Piezoelectric ultrasonic sensors use the piezoelectric effect of piezoelectric materials, such as quartz and tourmaline. The inverse piezoelectric effect converts high-frequency electrical vibrations into high-frequency mechanical vibrations to

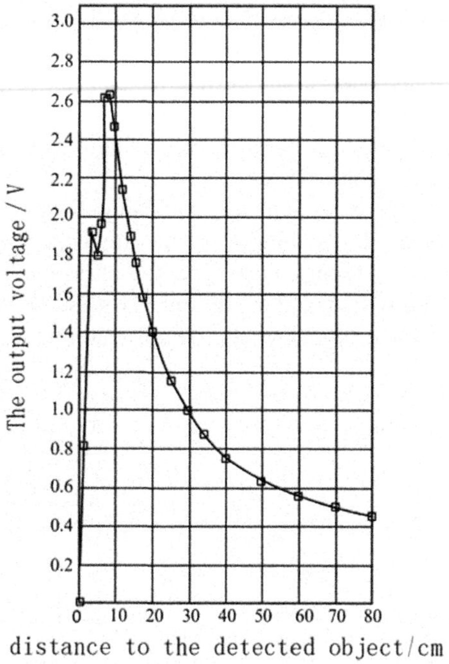

Fig. 3.20 Infrared sensor nonlinear output diagram

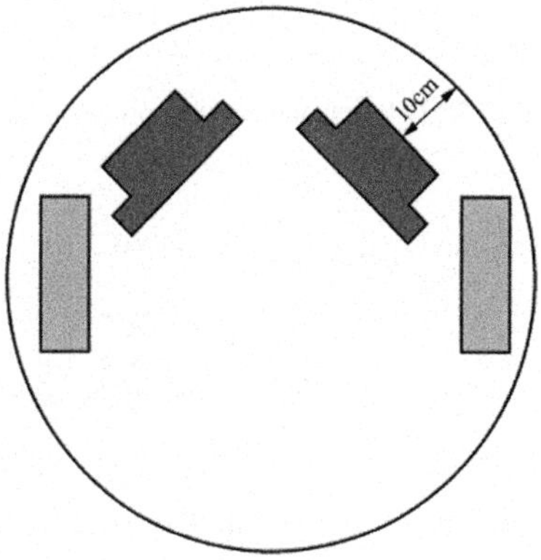

Fig. 3.21 Infrared sensor installation location

generate ultrasonic waves, which can be used as "transmitting" probes. Using the positive piezoelectric effect, the received ultrasonic vibration is converted into an electrical signal, which can be used as a "receiving" probe.

Due to different uses, piezoelectric ultrasonic sensors have various structural forms. Figure 3.22 shows one of them, the so-called dual probe (one probe transmits and the other probe receives). The piezoelectric wafer with the wafer seat is put into the metal shell, and the two sides of the piezoelectric wafer are plated with silver layers, which are used as electrode plates, the bottom surface is grounded, and the lead wires are connected above. The function of the damping block or absorption block is to reduce the mechanical quality factor of the piezoelectric sheet, absorb the sound energy, and prevent the piezoelectric sheet from continuing to vibrate due to inertia when the electric pulse oscillation stops. The best effect is achieved when the acoustic impedance of the damping block is equal to the acoustic impedance of the piezoelectric sheet.

There are two detection methods of ultrasonic distance sensor: pulse-echo type (Fig. 3.23) and FM-CW type (frequency modulation, continuous wave) (Fig. 3.24).

In the pulse-echo type, the ultrasonic wave is modulated with pulses and then transmitted, and the distance L of the measured object can be calculated according to the echo delay time Δt reflected by the measured object. Let the speed of sound

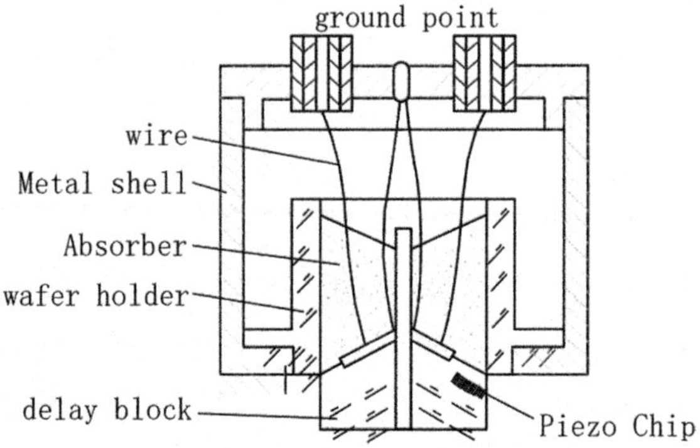

Fig. 3.22 Ultrasound dual-probe structure

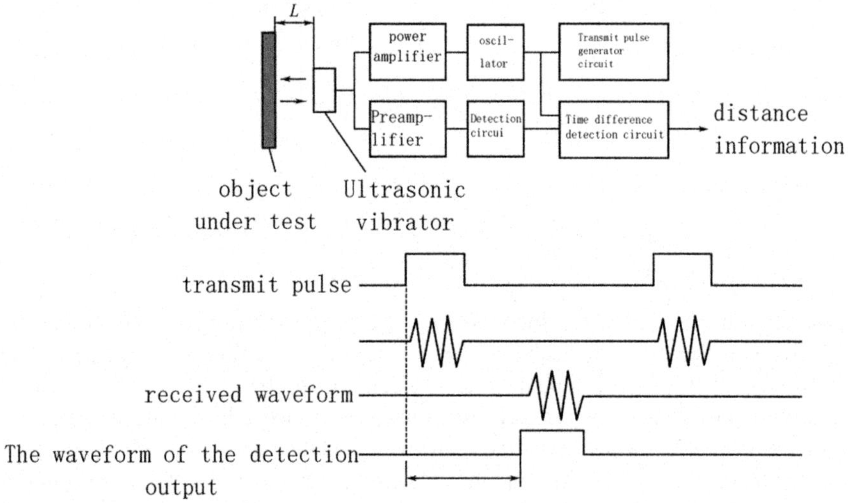

Fig. 3.23 Pulse echo detection principle

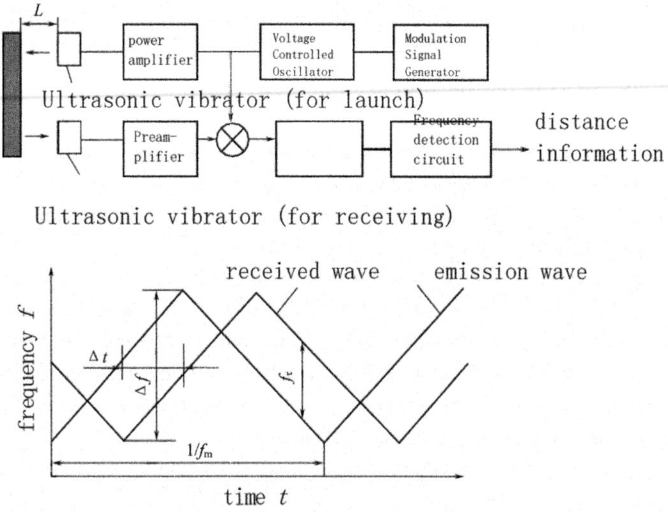

Fig. 3.24 FM-CW type ranging principle, f_τ —The frequency difference between the transmitted wave and the received wave; f_m —The frequency of the transmitted wave

in the air be v, if the air temperature is T°C, then the speed of sound is v = 331.5 + 0.607 T, and the distance between the measured object and the sensor is

$$L = v \cdot \Delta t/2 \tag{3.5}$$

The FM-CW method uses continuous waves to modulate the ultrasonic signal. The difference frequency $f\tau$ signal proportional to the distance L can be obtained by multiplying the received wave signal obtained by the measured object after delaying Δt time by the transmitted wave signal, and only taking out the low-frequency signal. Assuming that the frequency of the modulating signal is f_m, the bandwidth of the modulating frequency is Δf, and the distance between the measured object and the sensor is

$$L = \frac{f_\tau v}{4 f_m \Delta f} \tag{3.6}$$

Ultrasonic sensors have become standard in mobile robots, providing active detection tools on an inexpensive basis. In an ideal situation, the measurement precision of the ultrasonic sensor can obtain satisfactory results according to the above ranging principle. However, in a real environment, the accuracy and reliability of ultrasonic sensor data will decrease with the increase of distance and the complexity of environmental model. In general, the reliability of ultrasonic sensors is very low, and the results of ranging have great uncertainty, mainly in the following four points.

- The error of the distance measured by the ultrasonic sensor. In addition to the measurement precision of the sensor itself, it is also affected by changes in external conditions. For example, the propagation speed of sound waves in the air is greatly affected by temperature, and also has a certain relationship with air humidity.
- Ultrasonic sensor scattering angle. The sound wave emitted by the ultrasonic sensor has a scattering angle, and the ultrasonic sensor can sense that the obstacle is within the range of the fan-shaped area where the scattering angle is located, but cannot determine the exact position of the obstacle.
- Crosstalk. Robots are usually equipped with multiple ultrasonic sensors, and crosstalk problems may occur at this time, that is, the detection beam sent by one sensor is received by another sensor as its own detection beam. This situation usually occurs in a crowded environment, which can only be verified by repeated measurements in several different locations, and at the same time, the order of each ultrasonic sensor is reasonably arranged.
- The reflection of sound waves on the surface of an object. The imperfect reflection of acoustic signals in the environment is the biggest problem encountered by ultrasonic sensors in practical environments. When light, sound waves, electromagnetic waves, etc. hit a reflective object, any measured reflection retains only a part of the original signal, and the remaining part of the energy is either absorbed by the medium object, or scattered, or penetrates the object. Sometimes the ultrasonic sensor doesn't even pick up the reflected signal.

- Laser ranging sensor

Laser sensors are sensors that use laser technology to measure. It consists of a laser, a laser detector and a measuring circuit. Among them, a laser is a device that generates laser light. There are many types of lasers, which can be divided into solid lasers, gas lasers, liquid lasers and semiconductor lasers according to the working substance of the laser. Laser sensor is a new type of measuring instrument. Its advantages are that it can realize non-contact long-distance measurement, high speed, high precision, large measuring range, and strong anti-photoelectric interference ability.

Laser sensors can measure many physical quantities, such as length, speed, distance, etc. There are many types of laser ranging sensors. The principles of several commonly used laser ranging methods are described below, including pulsed laser ranging, phase laser ranging, and triangulation laser ranging.

The principle of the pulsed laser ranging sensor is: a pulsed laser with a very short duration is emitted by a pulsed laser, and after passing the distance to be measured, it hits the measured target, and a part of the energy will be reflected back, and the reflected pulsed laser is called an echo. The echo returns to the rangefinder, where it is received by the photodetector. According to the interval between the main wave signal and the echo signal, that is, the round-trip time difference between the laser pulse from the laser to the target to be measured, the distance to the target to be measured can be calculated.

Figure 3.25 shows the schematic diagram of the pulsed laser sensor ranging. When working, the laser emitting diode is aimed at the target to emit laser pulses. After

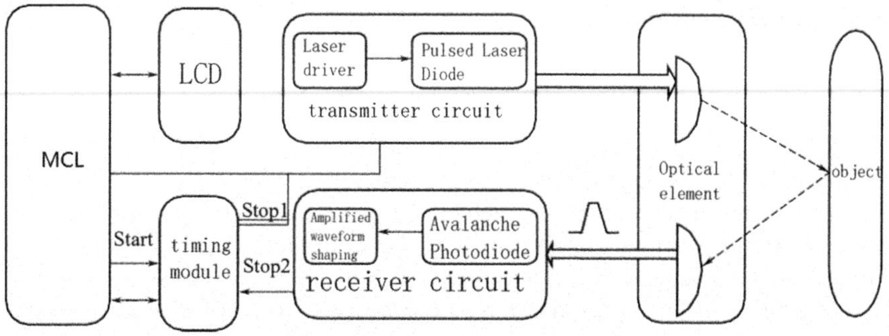

Fig. 3.25 Pulse laser sensor ranging principle

being reflected by the target, the laser light is scattered in all directions. Part of the scattered light returns to the sensor receiver, where it is picked up by the optical system and imaged onto the avalanche photodiode. The avalanche photodiode is an optical sensor with an internal amplification function, so it can detect extremely weak light signals and convert them into corresponding electrical signals.

If the time from when the light pulse is sent out to when it is returned and received is t, and the propagation speed of light is c, then the distance L between the laser sensor and the object to be measured can be obtained.

$$L = ct/2 \tag{3.7}$$

The principle of the phase laser ranging sensor is to modulate the intensity of the emitted laser light, and use the phase change of the modulated signal when the laser propagates in space to calculate the distance represented by the phase delay according to the wavelength of the modulated wave. That is, the indirect method of phase delay measurement is used to replace the direct measurement of the time required for the laser round-trip to achieve distance measurement. The precision of this method can reach the millimeter level.

The triangulation laser ranging sensor is the light emitted by the laser, which is focused by the convergent lens and then incident on the surface of the measured object. The receiving lens receives the scattered light from the incident light point and images it on the sensitive surface of the photoelectric position detector. When the object moves, the relative distance that the object moves is calculated by the displacement of the light spot on the imaging surface. The resolution of triangulation laser ranging is very high, which can reach the order of micrometers.

3.8 Smart Sensor

3.8.1 *Overview of Smart Sensor*

Smart Sensor (or Intelligent Sensor) was originally developed by NASA in 1978. A large number of sensors are required on the spacecraft to continuously send data information such as temperature, position, speed and attitude to the ground. It is difficult to use a large computer to process such complex data at the same time. If data is not lost and costs are reduced, there must be sensors that can achieve smart sensor integrated with computer. Smart sensors refer to sensors with functions of information detection, information processing, information memory, logical thinking and judgment. It not only has various functions of traditional sensors, but also has various functions such as data processing, fault diagnosis, nonlinear processing, self-correction, self-adjustment and human–computer communication. It is the product of the combination of microelectronic technology, microcomputer technology and detection technology.

In the early smart sensor, the output signal of the sensor is processed and converted, and then sent to the microprocessor part by the interface for calculation processing. In the 1980s, the smart sensor mainly took the microprocessor as the core, and integrated the sensor signal conditioning circuit, the microelectronic computer memorizer and the interface circuit into a single chip, so that the sensor had a certain artificial intelligence. In the 1990s, the smart measurement technology has been further improved, enabling the sensor to achieve miniaturization, structural integration, array type, digital type, easy to use and simple to operate, with self-diagnosis function, memory and information processing function, data storage function, multi-parameter measurement function, network communication function, logical thinking and judgment function.

Smart sensors are the main direction of the future development of sensor technology. In the future development, smart sensors will undoubtedly be further extended to research fields such as chemistry, electromagnetism, optics and nuclear physics.

- Definition of smart sensor

Smart sensor is a high-tech that is developing rapidly in today's world, and has not yet formed a standardized definition. In the early days, people simply and mechanically emphasized the close integration of the sensor and the microprocessor in the technology, and believed that " the sensitive element of the sensor and its signal conditioning circuit and the microprocessor are integrated on a single chip, which is a smart sensor ".

At present, the Chinese and English names of smart sensors have not been completely unified. The British call it "Intelligent Sensor"; the Americans are accustomed to calling it "Smart Sensor", the literal translation is "deft, smart sensor".

The so-called smart sensor is a sensor with a microprocessor, which has both information detection and information processing functions. The biggest feature of the smart sensor is to organically integrate the function of the sensor to detect information and the information processing function of the microprocessor. In a sense, it has an effect similar to human intelligence. It should be pointed out that the "with microprocessor" mentioned here includes two situations: one is that the sensor and the microprocessor are integrated on one chip to form a so-called "single-chip smart sensor"; the other is that the sensor can be equipped with a micro- processor. Obviously, the latter definition is broader, but both belong to the category of smart sensors.

- Composition of smart sensors

A smart sensor is composed of a combination of sensors and microprocessors. It makes full use of the computing and storage capabilities of the microprocessor to process the sensor data and adjust its internal behavior. Smart sensors have different names and uses depending on their sensing elements, and their hardware combinations are also different, but their structural modules are roughly similar, generally composed of the following parts:

- one or more sensitive devices;
- Microprocessor or microcontroller;
- Non-volatile erasable memory;
- Interface for bidirectional data communication;
- analog input and output interface (optional, such as A/D conversion, D/A conversion);
- Efficient power modules.

The microprocessor is the core of the smart sensor, it can not only calculate, store, and process the measured data of the sensor, but also adjust the sensor through the feedback loop. Because the microprocessor gives full play to the functions of various software, it can complete the tasks that are difficult for hardware to complete, thereby effectively reducing the difficulty of manufacturing, improving the performance of the sensor and reducing the cost. Figure 3.26 is a schematic diagram of a typical smart sensor structure.

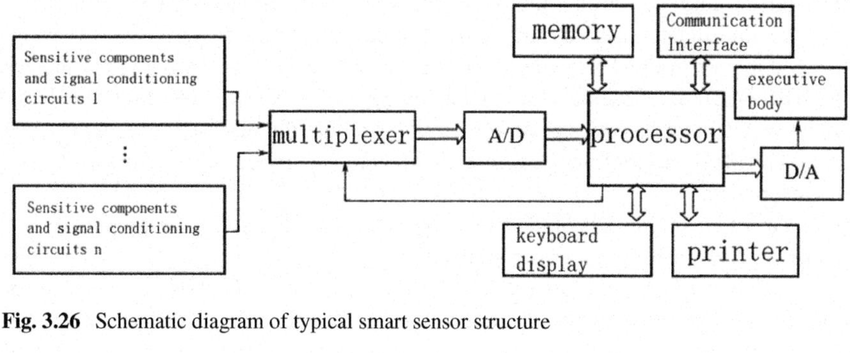

Fig. 3.26 Schematic diagram of typical smart sensor structure

The signal sensing devices of smart sensors often have two types: main sensor and auxiliary sensor. Taking an intelligent pressure sensor as an example, the main sensor is a pressure sensor that measures the measured pressure parameters, and the auxiliary sensors are a temperature sensor and an ambient pressure sensor. When the temperature sensor detects that the main sensor is working, the temperature of its pressure sensitive element changes due to the change of ambient temperature or the temperature of the measured medium, so as to correct and compensate the measurement error caused by temperature change according to its temperature change. Ambient pressure sensors measure changes in atmospheric pressure in the working environment to correct for their effects. The microcomputer hardware system amplifies, processes, stores and communicates with the computer the weak signal output by the sensor.

- Key technologies of smart sensors

Regardless of whether the smart sensor is a separate structure or an integrated structure, its intelligent core is a microprocessor, and many unique functions are realized on the basis of the least hardware by relying on powerful software advantages, while various software are directly related to their implementation principles and algorithms.

– Indirect sensing

Indirect sensing refers to the use of some process parameters or physical parameters that are easy to measure, by finding the relationship between these process parameters or physical parameters and the target measured variables that are difficult to directly detect, to establish a sensing mathematical model, using various calculation methods, using the software realizes the measurement of the variable to be measured. The core of indirect sensing of smart sensors lies in the establishment of a sensing model. The model can be established by the relevant principle equations of physics, chemistry and biology, or by the method of model identification. Different methods have their own advantages and disadvantages in application.

a. Modeling method based on process mechanism

The mechanism modeling method is based on a deep understanding of the process mechanism, and determine the mathematical relationship between difficult-to-measure dominant variables and easy-to-measure auxiliary variables by listing macroscopic or microscopic mass balances, energy balances, momentum balances, phase balance equations, and reaction kinetic equations. The model established based on the mechanism has strong interpretability and good extrapolation performance, and is an ideal indirect sensing model. Mechanism modeling has the following characteristics.

The mechanism models of the same object are very different in terms of structure and parameters, and the models are specific;

In the process of mechanism modeling, from the establishment of reaction intrinsic kinetics and various equipment models, the characterization of the heat and mass transfer effects of actual devices, to the estimation of a large number of parameters (from laboratory equipment to actual devices), each step is relatively complex;

The mechanism model is generally composed of algebraic equations, differential equations or partial differential equations. When the model structure is huge, the amount of calculation is large.

b. Based on data-driven modeling approach

For objects whose mechanism is not yet clear, a data-driven modeling method can be used to establish a soft-sensing model. This method extracts useful information from historical input and output data, and constructs the mathematical relationship between dominant and auxiliary variables. Data-driven modeling method is an important indirect sensing modeling method because it does not need to know much process knowledge. According to whether there is nonlinearity in the object, the modeling methods can be divided into linear regression modeling methods, artificial neural network modeling methods and fuzzy modeling methods. Linear regression modeling method is to collect a large number of measurement data of auxiliary variables and analysis data of dominant variables, and use statistical methods to extract the object information hidden in these data, so as to establish the mathematical model between dominant variables and auxiliary variables.

The artificial neural network modeling method directly models according to the input and output data of the object, uses the auxiliary variables that are easy to measure in the process as the input of the neural network, and uses the dominant variables as the output of the neural network, and solves the problem of indirect sensing modeling of the dominant variables through network learning. This method does not require prior knowledge of the object, and is widely used in system modeling where the mechanism is unclear and the nonlinearity is severe.

Fuzzy modeling is another effective tool for people to deal with complex system modeling, and it is also used in indirect sensing modeling, but the fuzzy neural network model that combines fuzzy technology and neural network is the most used.

c. Hybrid modeling approach

The limitations of mechanism-based modeling and data-driven modeling have led to the idea of hybrid modeling. For the process of simplified mechanism model, the simplified mechanism model and data-driven model can be combined to complement each other. The prior knowledge provided by the simplified mechanism model can save training samples for the data-driven model; the data-driven model can also compensate for the characteristics of the simplified mechanism model. Although the hybrid modeling method has good application prospects, its prerequisite is that there must be a simplified mechanism model.

It should be noted that the performance of the indirect sensing model is restricted by various factors, such as the selection of auxiliary variables, the transformation of sensor data, the preprocessing of sensor data, and the time series matching between the main and auxiliary variables.

– Linearization correction

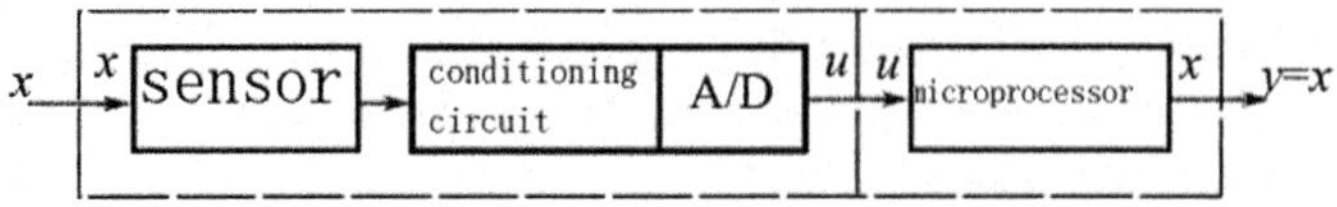

Fig. 3.27 Smart sensor linearization correction principle block diagram

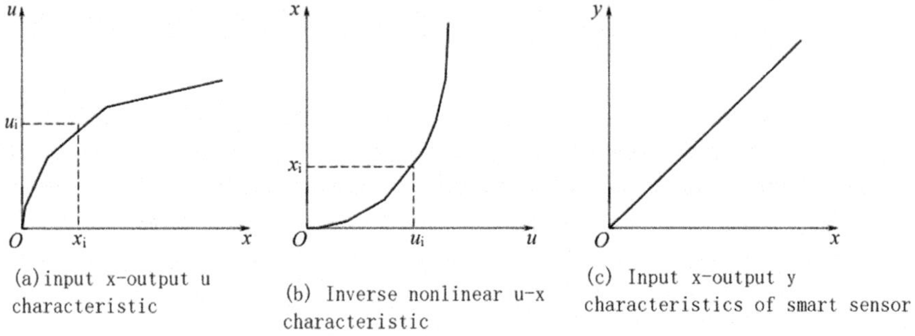

Fig. 3.28 Linearization of smart sensor input–output characteristics

The input physical quantity of an ideal sensor has a linear relationship with the converted signal. The higher the linearity, the higher the precision of the sensor. But in fact, the characteristic curve of most sensors has a certain nonlinear error.

Smart sensors can achieve linearization of sensor input–output. The outstanding advantage is that it is not limited to the nonlinear degree of the input–output characteristics of the front-end sensor, the conditioning circuit to the A/D conversion, and only requires good repeatability of the input x-output u characteristics. Figure 3.27 shows the principle block diagram of the linearization correction of the smart sensor. Among them, the input x-output u characteristic of the sensor and the conditioning circuit to the A/D converter is shown in Fig. 3.28a. The microprocessor performs anti nonlinear transformation on the input according to Fig. 3.28b, so that the input X and output y have a linear or approximately linear relationship, as shown in Fig. 3.28c.

At present, the nonlinear automatic correction methods mainly include look-up table method, curve fitting method and neural network method. Among them, the look-up table method is a piecewise linear interpolation method. Segment the nonlinear curve according to the accuracy requirements, and approximate the nonlinear curve with several broken lines. The neural network method uses a neural network to solve the undetermined coefficients of the inverse nonlinear characteristic fitting polynomial. The curve fitting method usually uses a row-order polynomial to approximate the inverse nonlinear curve, and each coefficient of the polynomial equation is determined by the least square method. The disadvantage of curve fitting method is that when there is noise, when using the principle of least square method to determine the undetermined coefficient, it may encounter ill conditioned conditions and cannot be solved.

- Self-diagnosis

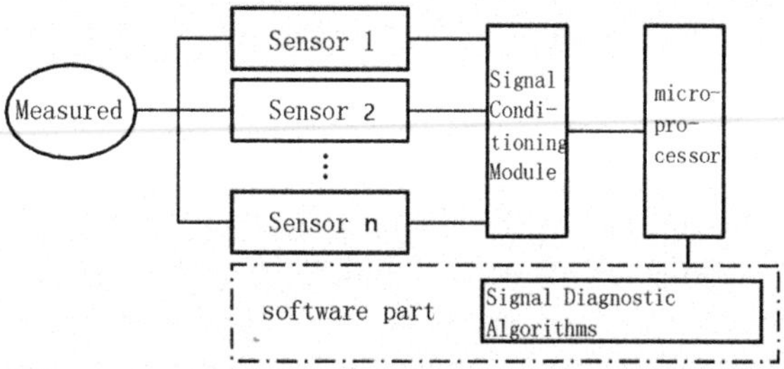

Fig. 3.29 Schematic diagram of hardware redundancy diagnosis method

The smart sensor self-diagnosis technology is commonly known as "self-test", which requires the detection of all parts of the smart sensor itself, including software resources and hardware resources, to verify whether the sensor can work normally and prompt relevant information.

Sensor fault diagnosis is one of the core contents of intelligent sensor self-test. The self-diagnostic program should judge whether the sensor is faulty, and realize fault location and identify fault types, so that corresponding countermeasures can be taken in subsequent operations. The fault diagnosis of the sensor is mainly based on the output of the sensor. Generally, there are hardware redundancy diagnosis method, mathematical model-based diagnosis method and signal processing-based diagnosis method.

a. Hardware redundancy diagnosis method

Redundant backup for sensors that are prone to failure, generally two, three or four identical sensors are used to measure the same measurand (Fig. 3.29), and the outputs of redundant sensors are compared with each other to verify the entire system output consistency. In general, the method uses two redundant sensors to diagnose the presence or absence of sensor faults, and three or more redundant sensors to separate the faulty sensors.

b. Diagnostic method based on mathematical model

Through a certain relationship between the measurement results or within the measurement result sequence, an appropriate mathematical model is established to characterize the characteristics of the measurement system, and whether there is a sensor failure is judged by comparing the difference between the model output and the actual output.

c. Diagnostic method based on signal processing

The detected signals are processed and exchanged directly to extract fault features, which avoids the difficulty of extracting mathematical models of objects in model-based methods. Although the diagnostic method based on signal processing is reliable, it also has limitations. For example, when some state divergence leads to output divergence, this method is not applicable; in addition, improper selection of threshold value will also cause false or missing reports of the method.

d. Fault diagnosis method based on artificial intelligence

Diagnosis method based on expert system stores the fault symptoms, fault modes, fault causes, handling opinions and other contents of an object in the knowledge base of the fault diagnosis expert system. Under the guidance of the reasoning organization, the expert system uses its knowledge to make inference and judgment according to the user's information, compares the observed phenomena with the potential causes, and forms the failure criterion. b. The diagnosis method based on neural network can use the powerful self-learning function, parallel processing ability and good fault tolerance ability of neural network. The neural network model is trained from the fault diagnosis case set of the diagnosis object, which avoids the real-time analysis of redundancy. modeling needs.

- Dynamic characteristic correction

When using the sensor to dynamically measure the transient signal, the sensor has a large dynamic error between the dynamic measurement result and the true value due to various reasons such as mechanical inertia, thermal inertia, electromagnetic energy storage element and circuit charge and discharge, that is, The variation curve of the output quantity with time is quite different from the variation curve of the measured value. Therefore, dynamic correction of the sensor is required.

In smart sensors, most of the methods for dynamically correcting the sensor are to connect the sensor with an additional correction link (Fig. 3.30), so that the synthesized total transfer function can reach an ideal or near-ideal (accuracy requirement) state. The main methods are:

a. Express dynamic characteristics of the sensor with low-order differential equations

The zero point of the transfer function of the compensation link is the same as the pole point of the sensor transfer function, and the dynamic compensation is realized by the method of zero-pole cancellation. This method requires the determination of the mathematical model of the sensor. The effectiveness of such dynamic compensators is limited due to simplifications and assumptions in determining the mathematical model.

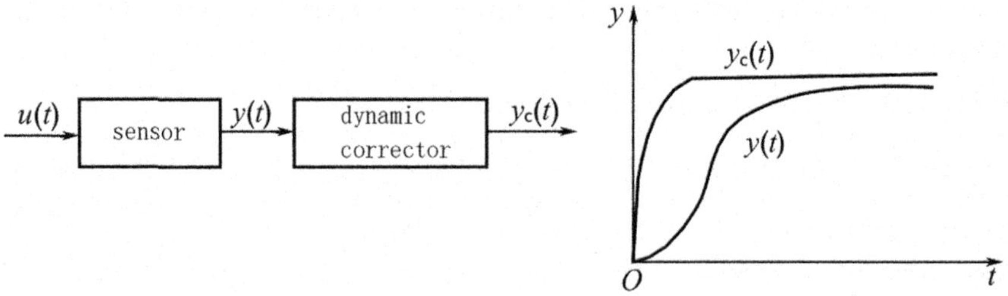

Fig. 3.30 Schematic diagram of dynamic correction principle

b. Establish a compensation link according to the actual characteristics of the sensor

According to the measured parameters of the sensor's response to the input signal and the output of the reference model, the dynamic compensation link is designed by the method of system identification. Due to the inevitable existence of various noises in the actual measurement system, there is a certain error in the dynamic compensation link of the identified sensor.

The core of adopting intermediate compensation and software correction for sensor characteristics is to correctly describe the data information and observation methods observed by the sensor, input and output models, and then determine the correction link.

– Self-calibration and adaptive range

a. Self-calibration

Self-calibration is somewhat equivalent to re-calibration before each measurement to remove systematic drift of the sensor. Self-calibration can adopt hardware self-calibration, software self-calibration and combination of software and hardware.

The self-calibration process of smart sensors is usually divided into the following three steps:

Zero calibration. Enter the zero-point standard value of the signal to perform zero-point calibration;

Calibration. Input signal standard value;

Measurement. Measure the input signal.

b. Adaptive range

The adaptive range of the smart sensor should comprehensively consider the range of the value to be measured, as well as the requirements for measurement accuracy, resolution and other factors to determine the setting of the gain (including attenuation) gears and determine the criteria for switching gears. Depends on the specific problem.

– Electromagnetic compatibility

The electromagnetic compatibility of the sensor refers to the adaptability of the sensor in the electromagnetic environment, that is, the ability to maintain its inherent performance and complete the specified function. It requires the sensor to be compatible with other electronic devices in the same space–time environment, neither being affected by electromagnetic interference nor affecting other electronic devices. Electromagnetic compatibility, as a performance indicator of smart sensors, has received more and more attention.

The electromagnetic interference of smart sensors includes the electromagnetic interference of the sensor itself (component noise, parasitic coupling, ground wire interference, etc.) and electromagnetic interference from outside the sensor (cosmic rays and lightning, external electrical and electronic equipment interference, etc.). Generally speaking, the suppression of sensor electromagnetic interference can be considered from several aspects of reducing noise signal energy, destroying the interference path, and improving its own anti-interference ability.

a. Electromagnetic shielding

Shielding is an effective way to suppress interference coupling. When the chip works at high frequency, the electromagnetic compatibility problem is very prominent. A better way is to shield the sensitive part with a shielding layer in the chip design, and connect the shielding layer of the chip to the shielding of the circuit. In the sensor, any place affected by electromagnetic field interference can be shielded to weaken the interference to ensure the normal operation of the sensor. Different shielding methods should be adopted for different interference fields, such as electrical shielding, magnetic shielding, electromagnetic shielding, and the shielding body should be well grounded.

b. Selection of components

The principle of derating is adopted and high-precision components are selected to reduce the thermal noise of the components themselves and reduce the internal interference of the sensor.

c. Ground

Grounding is an important measure to eliminate conducted interference coupling. When the signal frequency is lower than 1 MHz, the shielding layer should be grounded at one point. Because when multi-point grounding, the shielding layer forms a loop to the ground. If the potential of each grounding point is not completely equal, there will be an induced voltage, which is prone to inductive coupling, causing noise current in the shielding layer, which is coupled to the signal loop through distributed capacitance and distributed inductance.

d. Filtering

Filtering is one of the main means to eliminate conducted interference. Since the interfering signal has a different spectrum than the wanted signal, the filter can effectively suppress the interfering signal. The filtering methods for improving electromagnetic compatibility can be divided into hardware filtering and software filtering. π-type filtering is the recommended hardware filtering method on many standards. Software filtering relies on digital filters, which are unique to smart sensors to improve their ability to resist electromagnetic interference.

e. Reasonable design of the circuit board

The space where the sensor is located is often small, and most of it belongs to the near-field radiation. When designing, the area enclosed by the closed loop should be minimized to reduce parasitic coupling interference and radiation emission. In the case of high frequency, the distributed capacitance and inductance of printed circuit boards and components cannot be ignored.

3.8.2 Functions and Features of Smart Sensors

- Functions of smart sensors

Smart sensors mainly have the following functions:

- It has the functions of automatic zero adjustment, self-calibration and self-calibration. Smart sensors can not only automatically detect various measured parameters, but also perform automatic zero adjustment, automatic balance adjustment, and automatic calibration. Some smart sensors can also automatically complete the calibration work.
- It has the functions of logic judgment and information processing, and can perform signal conditioning or signal processing on the measured signal (preprocess and linearize the signal, or perform automatic compensation for parameters such as temperature and static pressure, etc.). For example, in an intelligent differential pressure sensor with temperature compensation and static pressure compensation, when the measured medium temperature and static pressure change, the compensation software of the intelligent sensor can automatically compensate according to a certain compensation algorithm to improve the measurement precision.
- It has self-diagnosis function. Through the self-checking software, the smart sensor can perform regular or irregular detection on the working status of the sensor and the system, diagnose the cause and location of the fault and make the necessary response, send a fault alarm signal, or display the operation prompt on the computer screen.
- It has configuration function and can be used flexibly. A variety of modular hardware and software can be set in the intelligent sensor system, and the user can issue instructions through the microprocessor to change the combined state of the hardware module and software module of the intelligent sensor to complete different measurement functions.
- It has data storage and memory functions, and can access test data at any time.
- It has two-way communication function and can directly communicate with microcomputers and other sensors and actuators through various standard bus interfaces and wireless protocols.

- Features of smart sensors

Compared with traditional sensors, smart sensors mainly have the following characteristics:

- High precision

The smart sensor has a number of functions to ensure its high precision, such as removing the zero point through automatic zero calibration, real-time comparison with the standard reference to automatically perform the overall system calibration, and automatically perform the correction of system errors such as the nonlinearity of the overall system. A large amount of data is statistically processed to eliminate the influence of accidental errors, so as to ensure that the measurement accuracy and resolution of the smart sensor are greatly improved.

- Wide range

Smart sensors have a wide measurement range and have a strong overload capability.

- High signal-to-noise ratio and high resolution

Because the smart sensor has the functions of data storage, memory and information processing, it can remove the noise in the input data and extract the useful signal by digital filtering, correlation analysis and other processing through software; through data fusion and neural network technology, the influence of cross-sensitivity in multi-parameter state can be eliminated, thereby ensuring the resolution of specific parameter measurement in multi-parameter state.

- Strong adaptive ability

The smart sensor has the functions of judgment, analysis and processing. It can decide the power supply of each part according to the working conditions of the system, and the data transmission rate of the high/upper computer, so that the system can work in the optimal low power consumption state and optimize the transmission efficiency.

- Cost-effective

The above-mentioned high performance of smart sensors is not obtained by pursuing the perfection of the sensor itself, meticulously designing and debugging all aspects of the sensor, and making "handicraft" like traditional sensor technology. Combined with microprocessors and microcomputers, it is realized by cheap integrated circuit technology and chips and powerful software, so it is cost-effective.

- Ultra-miniaturization and miniaturization

With the rapid promotion of microelectronics technology, smart sensors are developing in the direction of short, small, light and thin to meet the urgent needs of cutting-edge technologies in aviation, aerospace and national defense, at the same time, it also creates conditions for the miniaturized and portable development of general industrial and civil equipment.

- Low power consumption

Reducing power consumption has important implications for smart sensors. This not only simplifies the design of the system power supply and heat dissipation circuit, prolongs the service life of the smart sensor, but also creates favorable conditions for further improving the integration level of the smart sensor chip.

Smart sensors generally use large-scale or ultra-large-scale CMOS circuits, which greatly reduces the power consumption of the sensor, and some can be powered by laminated batteries or even button batteries. Standby mode can also be used to reduce the power consumption of smart sensors even when measurements are not being taken for a while.

3.8.3 Application of Smart Sensor in Robots

The application of intelligent sensor technology has made industrial robots a lot smarter, smart sensors add sense to robots and provide a foundation for intelligent robots to work with high precision and intelligence. The following introduces the smart sensors used in several intelligent robots.

- Two-dimensional vision smart sensor

The two-dimensional vision smart sensor is mainly a camera, which can complete the detection and positioning of object motion and other functions. Two-dimensional vision smart sensors have appeared for a long time. Many smart cameras can coordinate the action route of industrial robots. Adjust the robot's behavior based on the information it receives.

- Three-dimensional vision smart sensor

At present, three-dimensional (3D) vision smart sensors are gradually emerging. The 3D vision system must have two cameras to shoot at different angles, so that the 3D model of the object can be detected and recognized. Compared with 2D vision systems, 3D sensors can display things more intuitively.

- Force and torque intelligent sensor

The force-torque smart sensor is a smart sensor that can let the robot know the force. It can monitor the force on the smart robot arm, and based on the data analysis, it can guide the smart robot's next behavior.

- Collision detection smart sensor

The biggest requirement of industrial intelligent robots, especially collaborative robots, is safety. To create a safe working environment, intelligent robots must recognize what is unsafe. The use of a collision sensor allows the robot to understand what it hits and send a signal to pause or stop the robot's movement.

- Security smart sensor

Different from the collision detection sensors above, the use of safety sensors allows industrial robots to sense the presence of objects around them, and the presence of safety sensors prevents the robot from colliding with other objects.

- Other smart sensors

In addition to these, there are many other smart sensors, such as welding gap tracking sensors, in order to do a good job of welding, we need to be equipped with such a smart sensor, as well as tactile sensors and so on. Smart sensors bring a variety of senses to industrial robots that help them become smarter and work with greater precision.

3.9 Wireless Sensor Network Technology

With the promotion of automation technology, especially the requirements of the development of Fieldbus Control System (FCS), fieldbus networked intelligent sensors/transmitters with various communication modes have been developed.

With the progress and development of society, people put forward the networking requirements of sensor systems in a wider range of fields, such as multi-point remote monitoring of large machinery, multi-point monitoring in environmental areas, multi-point monitoring and remote consultation of critically ill patients, automatic real-time meter reading systems for electrical energy and distance teaching experiments, etc., the importance of wireless sensor networks is increasingly prominent.

Wireless Sensor Network (WSN) is a network composed of a large number of intelligent wireless sensor nodes that can communicate with each other according to a specific communication protocol. It integrates micro sensor technology, communication technology, embedded computing technology, distributed information processing technology as well as integrated circuit technology, it can collaboratively monitor, perceive and collect information of various environments or monitoring objects in the distribution area of the network in real time, and process and transmit this information. It has a wide range of application prospects in the fields of family and business.

The research on wireless sensor network started in the late 1990s. Since the last century, sensor networks have attracted great attention from academia, military and industry, and many research projects on wireless sensor networks have been launched in the United States and Europe. In particular, the United States has invested heavily in the research of sensor network technology through various channels such as the National Natural Science Foundation and the Ministry of Defense.

In China, research in the field of wireless sensor networks has also developed rapidly, and has been widely carried out in many research institutes and universities. The hotspots and difficulties of its research include: design of miniaturized node equipment; development of embedded real-time operating system suitable for sensor nodes; wireless sensor network architecture and various layer protocols; time synchronization mechanism and algorithm, sensor node's own positioning algorithm and the external target localization algorithm based on it and so on.

Especially after entering the twenty-first century, there are many novel solutions to the core problems of wireless sensor networks. However, this field is still in its infancy in general, and there are still many problems to be solved urgently.

3.9.1 Features of Wireless Sensor Networks

- System features

A wireless sensor network is a distributed sensor network. Its end is the sensor that can sense and check the external world. The sensors in the wireless sensor network communicate wirelessly, so the network settings are flexible, the location of the device can be changed at any time, and the connection to the internet can be wired or wireless.

Wireless sensor network is a self-organizing distributed network system composed of a large number of ubiquitous micro sensor nodes with wireless communication and computing capabilities. It is an "intelligent" system that can independently complete the specified tasks according to the environment. It has the ability to realize and control the behavior of swarm intelligence autonomous system, and can perceive cooperatively, collect and process the information of perceived objects in the geographical area covered by the network, and send it to the observer.

- Technical features

In the wireless sensor network system, a large number of sensor nodes are randomly deployed in or near the detection area, and these sensor nodes do not need personnel on duty. The nodes form a wireless network in a self-organizing manner to sense, collect and process specific information in the network coverage area in a cooperative manner, and can realize the collection, processing and analysis of information from any location at any time. The monitored data is transmitted back to the sink node through multi-hop relays along other sensor nodes, and finally the data in the entire area is transmitted to the remote control center for centralized processing by means of the convergence link. Users configure and manage the sensor network through management nodes, publish monitoring tasks and collect monitoring data.

At present, common wireless networks include mobile communication networks, wireless local area networks, bluetooth networks, Ad hoc networks, etc. Compared with these networks, wireless sensor networks have the following technical characteristics:

- The sensor node is small in size and has limited power supply capacity. Due to the limitation of price, volume and power consumption, the computing power, program space and memory space of nodes are much weaker than that of ordinary computers. This determines that the protocol layer should not be too complicated in the node operating system design. The network nodes are powered by batteries, and the capacity of the batteries is generally not very large. Some special application fields determine that the battery cannot be charged or replaced during use. Therefore, in the process of sensor network design, the use of any technology and protocol must be based on energy saving.
- The computing and storage capacity is limited. Due to the particularity of wireless sensor network applications, sensor nodes are required to have low price and low power consumption, which will inevitably lead to relatively weak processor capabilities and relatively small memory capacity. Therefore, how to use limited computing and storage resources to complete many collaborative tasks is also one of the challenges faced by wireless sensor network technology. In fact, with the improvement of low-power circuit and system design technology, many ultra-low-power microprocessors have been developed. At the same time, the general sensor node will also be equipped with some external memory. The current Flash

memory is a non-volatile storage medium that can operate at low voltage, write multiple times, and read infinitely.

- Centerless and self-organizing. There is no strict control center in wireless sensor network, all nodes have equal status, and it is a peer-to-peer network. Nodes can join or leave the network, and the failure of any node will not affect the operation of the entire network, and it has strong survivability. The deployment and expansion of the network does not need to rely on any preset network facilities. The nodes coordinate their respective behaviors through layered protocols and distributed algorithms. After the nodes are powered on, they can quickly and automatically form an independent network.
- The network is highly dynamic. A wireless sensor network is a dynamic network, and nodes can move anywhere; a node may exit the network due to battery exhaustion or other failures; a node may also be added to the network due to work needs. These will make the topology of the network change at any time, so the network should have dynamic topology organization.
- The number of sensor nodes is large and adaptive. The sensor nodes in the wireless sensor network are dense and the number is huge. In addition, the wireless sensor network can be distributed in a wide geographical area, the topology of the network changes rapidly, and once the network is formed, there is little human intervention, so the software and hardware of the wireless sensor network must have high robustness and fault tolerance, the corresponding communication protocol must be reconfigurable and adaptive.
- Multi-hop routing. The communication distance between nodes in the network is limited, generally within a range of several hundred meters, and a node can only communicate directly with its neighbors. If it wants to communicate with nodes outside its radio frequency coverage, it needs to route through intermediate nodes. The multi-hop routing of the fixed network is implemented by gateways and routers, while the multi-hop routing in the wireless sensor network is completed by ordinary network nodes without special routing equipment. In this way, each node can be both an initiator of information and a forwarder of information. Figure 3.31 is a schematic diagram of a multi-hop.

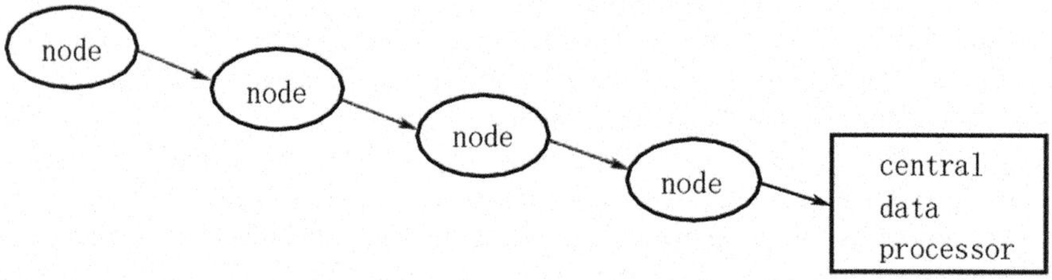

Fig. 3.31 A schematic diagram of a multi-hop

3.9.2 *Wireless Sensor Network Architecture*

- Network structure

Wireless sensor network is a multi-hop self-network system formed by a large number of inexpensive micro-sensors deployed in the monitoring area through wireless communication. Its purpose is to cooperatively perceive, collect and process the information of the perceived object in the network coverage area, and send it to the observer through the wireless network. Sensors, sensing objects and observers constitute the three elements of wireless sensor networks. The wireless sensor network architecture is shown in Fig. 3.32.

A wireless sensor network system usually includes a sensor node, a sink node, and a management node. A large number of sensor nodes are randomly deployed in or near the monitoring area (sensor field), which can form a network in a self-organizing manner. The data monitored by the sensor node is transmitted hop by hop along other sensor nodes. During the transmission process, the monitoring data may be processed by multiple nodes, routed to the sink node after multiple hops, and finally reach the management node through the internet or satellite. Users configure and manage the sensor network through management nodes, publish monitoring tasks and collect monitoring data.

The composition and function of wireless sensor network nodes include the following four basic units.

- Sensing unit: It is composed of sensor and analog/digital conversion function module. The sensor is responsible for collecting and converting the information of the sensing object.
- Processing unit: It is composed of embedded system, including CPU, memory, embedded operating system, etc. The processing unit is responsible for controlling the operation of the entire node, storing and processing the data collected by itself and the data sent by other nodes of the sensor.
- Communication unit: composed of wireless communication modules, wireless communication is responsible for realizing communication between sensor

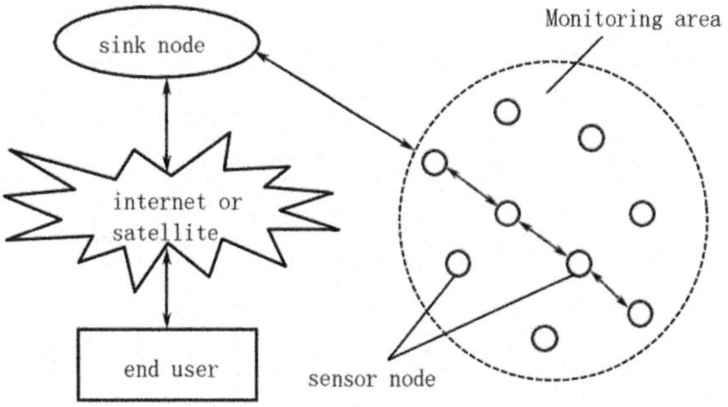

Fig. 3.32 Wireless sensor network architecture

nodes and between sensor nodes and user node management and control nodes, exchanging control messages and receiving/transmitting service data.
- Power supply part.

Most of the network nodes are powered by dry batteries or accumulators, and the capacity of the batteries is generally not large.

In addition, other functional units that can be selected include positioning systems, motion systems, and power generation devices.

- Node structure

A typical sensor network node is mainly composed of four parts: sensor module, processor module, wireless communication module and energy supply module, as shown in Fig. 3.33, where the dotted line represents optional modules. The sensor module is responsible for the collection and data conversion of information in the monitoring area; the processor module is responsible for controlling the operation of the entire sensor node, storing and processing the data collected by itself and the data sent by other nodes; the wireless communication module is responsible for wireless communication with other sensor network nodes, exchanging control information and sending and receiving collected data; the energy supply module provides the energy required for the operation of the sensor network nodes.

- Communication Architecture

The Open System Interconnect (OSI) network reference model has 7 layers, which are physical layer, data link layer, network layer, transport layer, session layer, presentation layer and application layer from bottom to top. Except for the physical layer and the application layer, all other layers communicate with the adjacent upper and lower layers. For example, the traditional wireless network and the existing Internet use a similar protocol layered design structure model, but some simplifications are made according to the optimization and combination of functions, and the three layers above the network layer are combined into a whole application layer. This simplifies the design of the protocol stack. Therefore, the Internet is a typical 5-layer

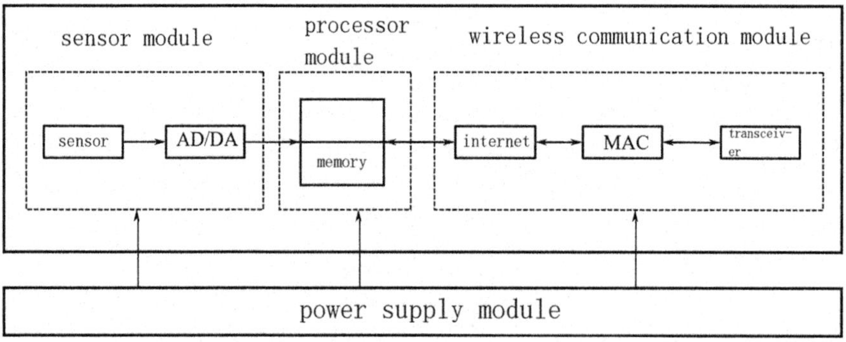

Fig. 3.33 The composition of wireless sensor nodes

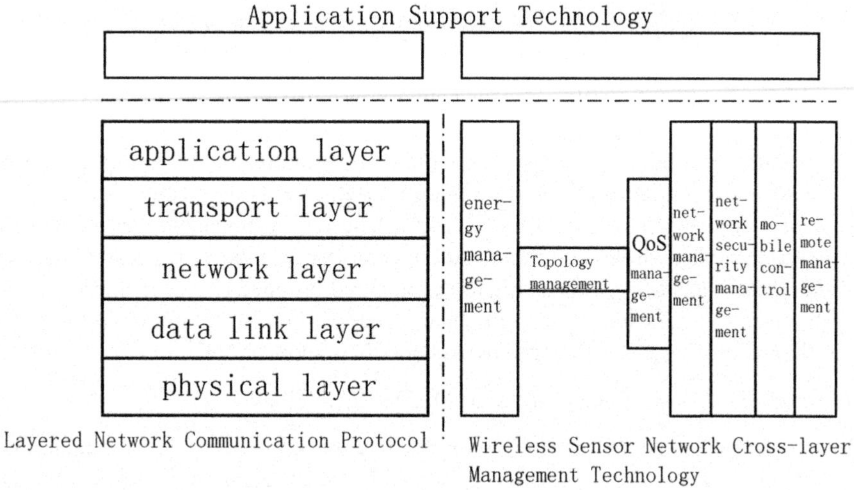

Fig. 3.34 Wireless sensor network protocol architecture

structure. The wireless sensor network protocol stack is also a 5-layer model, which corresponds to the physical layer, data link layer, network layer, transport layer and application layer of the OSI reference model. At the same time, the wireless sensor network protocol architecture defines cross-layer management technology and application support Technologies, such as energy management and topology management, are shown in Fig. 3.34.

The physical layer is responsible for sampling and quantizing the collected data, as well as signal modulation, transmission and reception, that is, the transmission of bit streams. Considering the presence of noise in the network environment and the movement of sensor nodes, the data link layer is mainly responsible for the multiplexing technology of data streams, data frame detection, medium access control, and error control, reducing the conflict of adjacent node broadcasts and ensuring reliable point-to-point, point-to-multipoint communication. The network layer maintains the data flow provided by the transport layer, mainly completes the route forwarding of data, and realizes the communication between sensors and sensors, sensors and information receiving centers. Routing technology is responsible for route generation and route selection. If the information is only transmitted within the wireless sensor, the transport layer may not be needed, but from the practical point of view, the wireless sensor network needs to communicate with the external network to transmit data. At this time, the transmission layer provides the conversion of the internal data-based addressing mode of the wireless sensor network to the addressing mode of the external network, that is, to complete the function of data format conversion. The application layer is composed of various sensor network application software systems, providing effective software development environment and software tools for users to develop various sensor network application software.

3.9.3 Key Technologies of Wireless Sensor Networks

The basic concept of wireless sensor network was first proposed 30 years ago. At that time, due to the limitations of technologies such as sensors, computers and wireless communication, this concept was only an imagination, and could not become a network technology that could be widely used, and its application was mainly limited to military systems. In recent years, with the advancement of microelectromechanical systems, wireless communication technology and low-cost manufacturing technology, it has become possible to develop and produce low-cost smart sensors with sensing, processing and communication capabilities, thus promoting the rapid development of wireless sensor networks and their applications.

- Micro Electro Mechanical System (MEMS) technology

Micro-electromechanical system technology is the key technology for manufacturing micro, low-cost, low-power sensor nodes. This technology is based on the micro machining technology of manufacturing micron mechanical parts. Through the use of highly integrated processes, various electromechanical components and complex MEMS can be manufactured. There are different types of micromachining techniques, such as plane machining, batch machining, surface machining, etc., which use different processing procedures. Most micromachining operations are performed on a 10–100 μm thick substrate composed of silicon, crystalline semiconductors or quartz crystals to complete a series of processing steps, such as film decomposition, photolithography, surface etching, oxidation, electroplating, wafer bonding, etc., different processing steps can have different processing steps. By integrating different components on a single substrate, the size of the sensor node can be greatly reduced. Using MEMS technology, many components of sensor nodes, such as sensors, communication modules and power supply units, can be miniaturized, and the cost and power consumption of nodes can also be greatly reduced through mass production.

- Wireless communication technology

Wireless communication technology is the key technology to ensure the normal operation of wireless sensor network. In the past decades, wireless communication technology has been extensively studied in the field of traditional wireless networks, and significant progress has been made in various aspects. At the physical layer, various modulation, synchronization, and antenna technologies have been designed for different network environments to meet different application requirements. At the link layer, network layer and higher layers, various efficient communication protocols have been developed to solve various network problems, such as channel access control, quality of service, network security, etc. These technologies and protocols provide a rich technical basis for the design of wireless communication in wireless sensor networks.

At present, most traditional wireless networks use radio frequency (RF) for communication, including microwave and millimeter waves. The main reason is that RF communication does not require line of sight transmission and can provide omnidirectional connections. However, radio frequency communication also has some limitations, such as large radiation and low transmission efficiency, so it is not the best transmission medium for micro, energy-limited sensor communication. Optical Radio Communication is another transmission medium that may be suitable for sensor network communication. Compared with radio frequency communication, wireless optical communication has many advantages. For example, optical transmitters can be made very small; optical signal transmission can obtain large antenna gain, thereby improving transmission efficiency; optical communication has strong directivity, enabling it to use Spatial Division Multiple Access (SDMA), reduces communication overhead, and has the potential to achieve higher energy efficiency than the multiple-access approach used in RF communications. However, optical communication requires line-of-sight transmission, which limits its application in many sensor networks.

For traditional wireless networks (such as cellular communication systems, wireless local area networks, mobile ad hoc networks, etc.), most of the communication protocols are designed without considering the special problems of wireless sensor networks, so they cannot be directly used in sensor networks. In order to solve various unique network problems in wireless sensor networks, the characteristics of wireless sensor networks must be fully considered in the design of communication protocols.

3.9.4 Hardware and Software Platform

The development of wireless sensor networks largely depends on the ability to research and develop low-cost, low-power hardware and software platforms suitable for sensor networks. Using MEMS technology, the size and cost of sensor nodes can be greatly reduced and reduced. In order to reduce the power consumption of nodes, energy-sensing technology and low-power circuit and system design technology can be used in hardware design. At the same time, dynamic power management (DPM) technology can also be used to efficiently manage various system resources and further reduce the power consumption of nodes. For example, when the node has little or no load to process, all idle components can be dynamically shut down or put into a low-power sleep state, thereby greatly reducing the power consumption of the node. On the other hand, if energy sensing technology is adopted in the design of system software, the energy efficiency of nodes can also be greatly improved. The system software of sensor nodes mainly includes operating system, network protocol and application protocol. In the operating system, the task scheduler is responsible for scheduling various tasks of the system under certain time constraints. If energy sensing technology is adopted in the task scheduling process, it will be able to effectively prolong the life of sensor nodes.

At present, the development of many low-power sensor hardware and software platforms adopts low-power circuit and system design technology and power management technology. The emergence and commercialization of these platforms further promote the application and development of wireless sensor networks.

- Hardware platform

The hardware platforms of sensor nodes can be divided into three categories: enhanced general-purpose personal computers, dedicated sensor nodes, and sensor nodes based on System-on-Chip (SoC).

– Enhanced general-purpose personal computer

Such platforms include various low-power embedded personal computers (such as PCI04) and personal digital assistants (PDAs), which typically run operating systems already on the market (such as Win CE or Linux) and use standard wireless communication protocol (such as IEEE 802.11 or Bluetooth). Compared to dedicated sensor nodes and system-on-a-chip sensor nodes, these PC-like platforms have greater computing power, enabling the inclusion of richer network protocols, programming languages, middleware, application programming interfaces (APIs), and other software.

– Dedicated sensor nodes

Such platforms include Berkeley motes, UCLA Medusa and MIT μAmp and other series. These platforms usually use existing chips on the market, which have the characteristics of small waveform factor, low power consumption of calculation and communication, simple sensor interface and so on.

– Sensor nodes based on SoC

Such platforms include Smart Dust, etc., which are based on CMOS, MEMS and RF technologies and aim to achieve ultra-low power consumption and small footprint, and have certain sensing, computing and communication capabilities.

Among all the above platforms, Berkeley Motes has been widely used in the field of sensor network research due to its small form factor, open source code and commercialization.

- Software platform

A software platform can be an operating system that provides various services, including file management, memory allocation, task scheduling, peripheral drivers, and networking, or a language platform that provides a component library for programmers. Typical sensor software platforms include TinyOS, nesC, TinyGALS, etc. TinyOS is one of the earliest operating systems to support sensor network applications on resource-constrained hardware platforms such as Berkeley Motes. This operating system is event-driven, uses only 178 bytes of memory, but is able to support function such as communication, multitasking, and code modularization. nesC is an extension of the C language to support the design of TinyOS, providing a set of language components and constraints for implementing TinyOS components and applications. TinyGALS is a language for TinyOS that provides a way to concurrently execute multiple component threads driven by events. Unlike nesC, TinyGALS solves concurrency problems at the system level rather than at the component level.

3.9.5 Interconnection of Wireless Sensor Network and Internet

In most cases, wireless sensor networks work independently. For some important applications, it is very necessary to connect the wireless sensor network to other networks. For example, in disaster monitoring applications, a sensor network deployed in a disaster area with harsh environments is connected to the Internet. The sensor network can transmit data to a gateway through a satellite link, and the gateway is connected to the Internet, which enables the monitoring personnel to obtain the real-time data in the disaster area. There are two main solutions to solve the interconnection between wireless sensor network and Internet: homogeneous network and heterogeneous network.

Homogeneous network refers to a network in which one or more independent gateway nodes are set between the wireless sensor network and the Internet to realize the access of the wireless sensor network to the Internet. Except for the gateway node, all nodes have the same resources. The main idea of this structure is: use the gateway to shield the sensor network and provide real-time information services and interoperability functions to remote Internet users. The network places the interface with the Internet standard IP protocol at the gateway node outside the wireless sensor network. This is more in line with the data flow mode of the wireless sensor network, easy to manage, and does not need to make major adjustments to the wireless sensor network itself; there is no need to adjust the sensor network itself. The disadvantage of this structure is that a large number of data flows are gathered around the nodes close to the gateway, so that the energy consumption of the nodes near the gateway is too fast, and the energy consumption in the network is unevenly distributed, thus reducing the lifetime of the sensor network. Figure 3.35 shows the interconnection structure of a homogeneous network.

If some nodes in the network have stronger capabilities than most other nodes and are given IP addresses, these interface nodes can implement the TCP/IP protocol for the Internet and a specific transmission protocol for the sensor network. for heterogeneous networks, this kind of network is called heterogeneous network. The main idea of this structure is to use specific nodes to shield the sensor network and provide real-time information services and interoperability functions to remote Internet users. To balance the load within the sensor network, multiple pipes can be established between these interface nodes. The characteristic of the heterogeneous network is that some nodes with high energy are assigned IP addresses as the interface with the internet standard IP protocol. These high-capacity nodes can complete complex tasks and take on more loads. The difficulty lies in the inability to have a

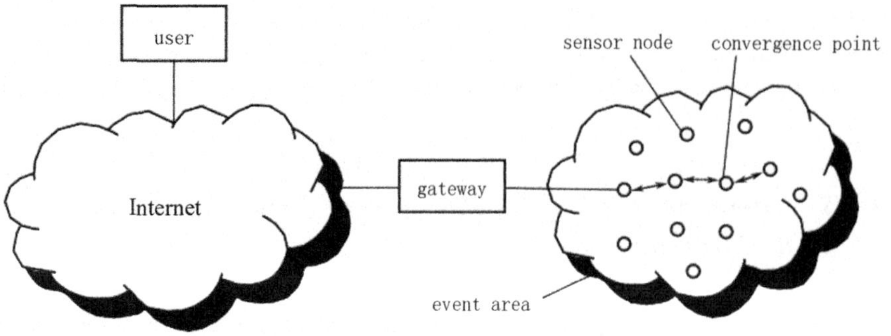

Fig. 3.35 Homogeneous network interconnection with a single gateway

clear definition of the so-called "high-capacity" of nodes. Figure 3.36 Heterogeneous network interconnection using interface nodes.

Compared with the homogeneous network interconnection, the heterogeneous network interconnection has a more uniform energy consumption distribution, and can better integrate the data flow within the sensor network, thereby reducing information redundancy. However, the interconnection of heterogeneous networks needs to adjust the routing and transmission protocols of sensor networks to a large extent, which increases the complexity of designing and managing sensor networks.

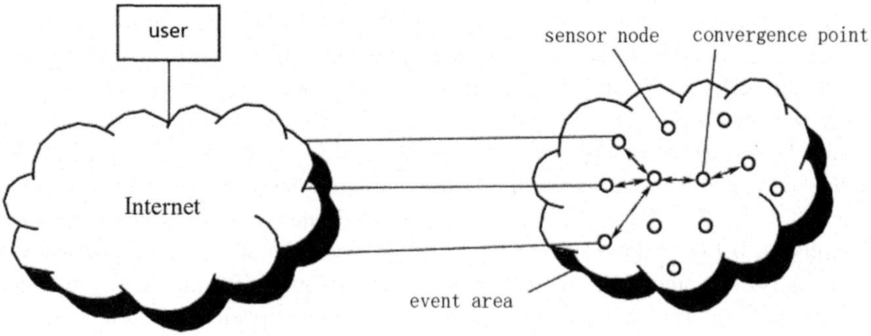

Fig. 3.36 Heterogeneous network interconnection using interface nodes

Chapter 4
Vision-Based Mobile Robot Positioning Technology

4.1 Mobile Robot Vision System

4.1.1 Basic Concepts of Robot Vision

The word "vision" is first and foremost a biological concept. In addition to the narrow concept of "light acting on the visual organs of living beings", its broad definition also includes the processing and recognition of visual signals, that is, the use of the optic nervous system and the brain center to perceive the size, color, brightness and darkness, orientation and other abstract information of external objects through visual signals. Vision is an important way for humans to obtain information from the outside world. According to statistics, 80% of the information obtained by human sense organs is obtained by vision. In order to make robots more intelligent and adapt to various complex environments, vision technology is introduced into robotics. Through the vision function, the robot can realize product quality inspection, target recognition, positioning, tracking, autonomous navigation and other functions.

With the development of science and technology, especially the development of computer science and technology, automation technology, pattern recognition and other disciplines. As well as the practical needs of application fields such as autonomous robots and industrial automation, it has become particularly important to endow these intelligent machines with human vision capabilities, and thus form a new discipline—machine vision. For ease of understanding, some concepts in machine vision are briefly introduced below.

Camera Calibration: It is the process of obtaining the internal parameters and external parameters of the camera. Usually, the internal parameters of the camera are also called intrinsic parameters, which mainly include the image coordinates of the center point of the optical axis, the magnification coefficient from the imaging plane coordinates to the image coordinates (also known as the focal length normalization coefficient), the lens distortion coefficient, etc.; The external parameters of the camera, also known as extrinsic parameter, are the representation of the camera

T. Guo et al., *Special Robot Technology*, Advanced and Intelligent Manufacturing in China, https://doi.org/10.1007/978-981-99-0589-8_4

coordinate system in the reference coordinate system, that is, the transformation matrix between the camera coordinate system and the reference coordinate system.

Vision System Calibration: The determination of the relationship between the camera and the robot is called vision system calibration. For example, the calibration of the hand-eye system is to obtain the relationship between the camera coordinate system and the robot coordinate system.

Planar Vision: A vision system that only measures the information of the target in the plane, which is called a plane vision system. Planar vision can measure the two-dimensional position information of the target and the one-dimensional pose of the target. Plane vision generally uses a camera, and the calibration of the camera is relatively simple.

Stereo Vision: A vision system that measures the information of a target in a three-dimensional Cartesian space, called a stereo vision system. Stereo vision can measure the three-dimensional position information of the target, as well as the three-dimensional attitude of the target. Stereo vision generally uses two cameras, and the internal and external parameters of the cameras need to be calibrated.

Active Vision: A vision system that actively illuminates the target or actively changes the camera parameters, which is called an active vision system. Active vision can be divided into structured light active vision and variable parameter active vision.

Passive Vision: Passive vision uses natural measurements, such as binocular vision, which is passive vision.

Vision Measure: The measurement of the position and attitude of the target based on the visual information obtained by the camera is called vision measurement.

Vision Control: The position and attitude of the target are obtained according to the visual measurement, and it is used as a given or feedback to control the position and attitude of the robot, which is called vision control. In short, the so-called visual control is the control of the robot based on the visual information obtained by the camera. In addition to the usual position and posture, visual information also includes the color, shape, size, etc. of the object.

4.1.2 Main Application Fields of Mobile Robot Vision System

Vision is the most abundant means for humans to obtain information. Usually, most of the information of humans comes from the eyes, while for drivers, more than 90% of the information comes from vision. Similarly, the vision system is one of the important components of the mobile robot system, and the vision sensor is also the perception device for the mobile robot to obtain surrounding information. In the past ten years, as researchers have invested a lot of research work, computer vision, machine vision and other theories have been continuously developed and improved. The vision system of mobile robots has involved image acquisition, compression coding and transmission, image enhancement, edge detection, threshold segmentation, target recognition, 3D reconstruction, etc., covering almost all aspects of machine vision. At present, the mobile robot vision system is mainly used in the following three aspects:

- Visual inspection of products is used instead of human visual inspection. Including: shape inspection, that is, checking and measuring the geometric size, shape and position; defect inspection, that is, checking whether the part is damaged or scratched; complete inspection, that is, checking whether the parts on the part are complete.
- Identify the components to be assembled one by one, determine their spatial position and direction, guide the robot's hand to accurately grasp the required parts, and place them in the designated position to complete the tasks of classification, handling and assembly.
- Navigate the mobile robot, use the vision system to provide the mobile robot with external information of its environment, so that the robot can autonomously plan its route, avoid obstacles, safely reach the destination and complete the designated work tasks.

The mobile robot is endowed with human visual function, and can process video like a human, so that it has the ability to obtain information from the external environment. Work is extremely important. The vision system includes both hardware and software. The former lays the foundation of the system, while the latter is usually indispensable. It includes the algorithm of image processing and the interface program of human–computer interaction.

4.1.3 Mobile Robot Monocular Vision System

Video cameras can be divided into analog video cameras and digital video cameras. Analog cameras, also known as TV cameras, are used less often these days. Digital video cameras are commonly used CCD (Charge Couple Device) cameras, which are often referred to now as cameras. From the perspective of mobile robot vision technology, cameras can be divided into three categories: monocular, binocular, and panoramic. The camera is usually represented by a model. For a monocular camera, the simplest pinhole model is generally used.

- Camera reference coordinate system

In order to describe the optical imaging process, the following coordinate systems are involved in the computer vision system, as shown in Fig. 4.1.

- Pixel Coordinate System. Represents the projection of 3D points in the scene onto the image plane. The digital image collected by the camera can be stored as an array in the computer, and the value of each element (Pixel) in the array is the brightness (grayscale) of the image point. As shown in Fig. 4.1, define a rectangular coordinate system u-v on the image, the origin of which is at the upper left corner of the CCD image plane, the axis u is parallel to the CCD image plane

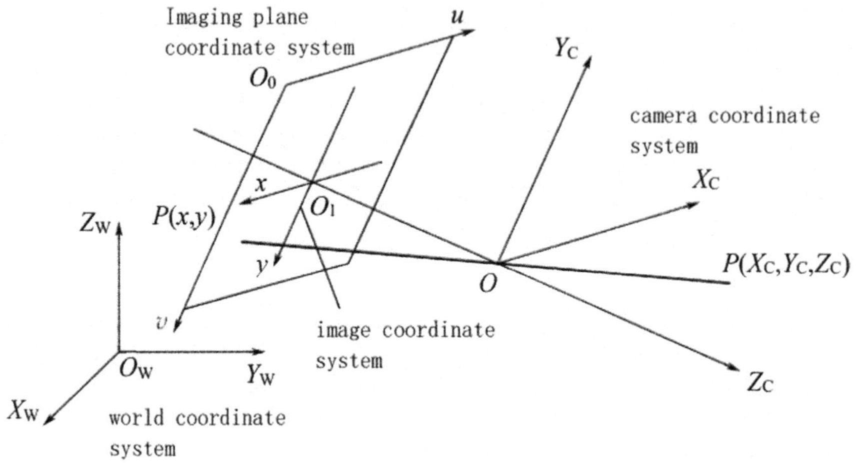

Fig. 4.1 Camera coordinate system

horizontally to the right, the axis v is perpendicular to the axis u and vertically downward, each pixel The coordinates (u, v) are the number of columns and rows of the pixel in the array, respectively. Therefore (u, v) is the coordinate of the image coordinate system in pixels.

- Retinal Coordinate System. Retinal coordinate system also known as the image physical coordinate system. Since the image pixel coordinate system only represents the position of the pixel, and the pixel has no actual physical meaning. Therefore, it is necessary to establish a plane coordinate system with physical units such as millimeters. In the x–y coordinate system, the origin O_1 is defined at the intersection of the optical axis of the camera and the image plane, which is called the principal point of the image. That is, the origin of the coordinates is at the center of the CCD image plane (u0, v0), the axis X and Y are respectively parallel to the coordinate axes of the image coordinate system, and the coordinates are represented by (x, y). This point is generally at the center of the image, but may be slightly off due to camera production.
- Camera Coordinate System. Taking the optical center of the camera as the origin of the coordinate system, the axis X_c and the axis Y_c are parallel to the axis of the imaging coordinate system, and the axis Z_c is the optical axis of the camera, which is perpendicular to the image plane; the intersection of the optical axis and the imaging plane is called the image principal point, The coordinate system satisfies the right-hand rule. The scene points are represented in the form of observer-centric data, represented by (X_c, Y_c, Z_c). The cartesian coordinate system composed of point P and axis X_c, Y_c, Z_c is called the camera coordinate system; OO_1 is the focal length of the camera.
- World Coordinate System. In the environment, a reference coordinate system can also be selected to describe the position of the camera and the object. This coordinate system is called the world coordinate system, also called the real coordinate

system or the objective coordinate system. It is used to represent the absolute coordinates of the scene point, represented by (X_w, Y_w, Z_w).

- Camera model

– Pinhole model

Perspective projection is the most commonly used imaging model, which can be approximated by pinhole perspective or central perspective projection models. The pinhole model is the simplest of various camera models. It is an approximate linear model of the camera. It actually only includes perspective projection transformation and rigid body transformation, and does not include camera distortion factors, but it is the basics of other models. The pinhole model is characterized by the fact that all rays from the scene pass through a projection center, which corresponds to the center of the lens. The line passing through the center of projection and perpendicular to the image plane is called the projection axis or optical axis. The casting projection produces an upside-down image, sometimes an upright virtual plane equidistant from the actual imaging plane to the pinhole is envisaged. where $x-y-z$ is a rectangular coordinate system fixed on the camera, following the right-hand rule, its origin is at the center of the projection, the z-axis coincides with the projection and points to the scene, the X_c-axis, Y_c-axis are parallel to the image plane's coordinate axes x and y, The distance OO_1 between the plane $Xc-Yc$ and the image plane is the focal length f of the camera.

In the relationship between image coordinates and physical coordinates described in Fig. 4.1, O_0 is the origin of the image coordinate system, and the coordinates of point P in the image pixel coordinate system are (u, v). Suppose (u_0, v_0) represents the coordinates of O_1 in the u-v coordinate system, and d_x and d_y represent the physical dimensions of each pixel on the horizontal axis x and vertical axis y, respectively. Without considering the distortion, any pixel in the image is in the relationship between the image coordinate system and the pixel coordinate system is as follows:

$$\begin{aligned} u &= \frac{x}{dx} + u_0 \\ v &= \frac{y}{dy} + v_0 \end{aligned} \tag{4.1}$$

Assuming the units in the physical coordinate system are millimeters, the units of dx are millimeters per pixel. Then the unit of x/dx is the pixel, which is the same as the unit of u. For convenience, the above formula is expressed in matrix form as:

$$\begin{bmatrix} u \\ v \\ 1 \end{bmatrix} = \begin{bmatrix} 1/dx & 0 & u_0 \\ 0 & 1/dy & v_0 \\ 0 & 0 & 1 \end{bmatrix} \begin{bmatrix} x \\ y \\ 1 \end{bmatrix} \tag{4.2}$$

The world coordinate system is introduced to describe the position of the camera, which describes the location relationship between the camera and the object in a three-dimensional environment. The coordinate system consists of the axis X_w, Y_w and Z_w. A rotation in any dimension can be expressed as the product of a coordinate vector and a suitable square matrix. The relationship between the camera coordinate system and the world coordinate system can be described by a rotation matrix R and a translation vector T. Therefore, if the homogeneous coordinates of a point P in the known space in the world coordinate system and the camera coordinate system are $(Xw, Yw, Zw, 1)^T$ and $(Xc, Yc, Zc, 1)^T$ respectively, then:

$$\begin{bmatrix} X_c \\ Y_c \\ Z_c \end{bmatrix} = R \begin{bmatrix} X_w \\ Y_w \\ Z_w \end{bmatrix} + T \tag{4.3}$$

Among them, R is the orthogonal unit rotation matrix, T is the three-dimensional translation vector. For any point P in space, the relationship between the camera coordinate system and the image coordinate system can be written as:

$$x = f\frac{Xc}{Zc}, y = f\frac{Yc}{Zc} \tag{4.4}$$

From formulas (4.2)–(4.4) we get:

$$Z_c \begin{bmatrix} u \\ v \\ 1 \end{bmatrix} = \begin{bmatrix} f/dx & s & u_0 \\ 0 & f/dy & v_0 \\ 0 & 0 & 1 \end{bmatrix} \begin{bmatrix} R & T \end{bmatrix} \begin{bmatrix} X_w \\ Y_w \\ Z_w \\ 1 \end{bmatrix}$$

$$= \begin{bmatrix} k_u & s & u_0 \\ 0 & k_v & v_0 \\ 0 & 0 & 1 \end{bmatrix} \begin{bmatrix} R & T \end{bmatrix} \begin{bmatrix} X_w \\ Y_w \\ Z_w \\ 1 \end{bmatrix}$$

$$= K\begin{bmatrix} R & T \end{bmatrix} \begin{bmatrix} X_w \\ Y_w \\ Z_w \\ 1 \end{bmatrix} = P \begin{bmatrix} X_w \\ Y_w \\ Z_w \\ 1 \end{bmatrix} \tag{4.5}$$

In the formula, P is a 3*4 matrix, called projection matrix; s called torsion factor, $k_u = f/dx, k_v = f/dx$; K is completely determined by k_u, k_v, s, u_0, v_0. K is only related to the internal structure of the camera, called the camera internal parameter matrix; [R T] is determined by the orientation of the camera relative to the world coordinate system, called camera extrinsic parameters.

– Distortion model

Since the pinhole model is only an approximation of the actual camera model, and there are various distortions and deformations of lens, the imaging of the actual camera is much more complicated. After introducing different deformation corrections, various nonlinear imaging models are formed.

There are three main types of lens distortion: radial distortion, centrifugal distortion, and thin prism distortion. The radial distortion only causes the radial position deviation of the image point, while the centrifugal distortion and thin prism distortion will cause both the radial position deviation and the tangential position deviation of image points.

a. Radial distortion: The radial distortion is mainly caused by the lens shape defect, which is symmetrical about the main optical axis of the camera lens. Positive distortion is also called pincushion distortion, and negative distortion is called barrel distortion. The mathematical model of radial distortion is:

$$\Delta_r = k_1 r^3 + k_2 r^5 + k_3 r^7 + \cdots \tag{4.6}$$

In the formula, $r = \sqrt{u_d^2 + v_d^2}$, is the distance from the image point to the center of the image, k_1, k_2, k_3, ... is the radial distortion coefficient.

- Centrifugal distortion: it is caused by the inconsistency between the optical center and the geometric center of the optical system. This type of distortion includes both radial distortion and tangential distortion that is asymmetrical to the main optical axis of the camera lens, and its rectangular coordinates are in the form of:

$$\begin{cases} \Delta_{ud} = 2p_1 u_d v_d + p_2(u_d^2 + 3v_d^2) + \cdots \\ \Delta_{vd} = p_1(3u_d^2 + v_d^2) + 2p_2 u_d v_d + \cdots \end{cases} \tag{4.7}$$

  In the formula, p_1, p_2 is the tangential distortion coefficient.

- Thin prism distortion: It is caused by lens design defects and processing and installation errors, such as a small inclination between the lens and the camera imaging surface. This kind of distortion is equivalent to adding a thin prism to the optical system, which causes not only radial position deviation, but also tangential position deviation. Its rectangular coordinate form is:

$$\begin{cases} \Delta_{up} = s_1(u_d^2 + v_d^2) + \cdots \\ \Delta_{vp} = s_2(u_d^2 + v_d^2) + \cdots \end{cases} \tag{4.8}$$

where s_1 and s_2 are the thin prism distortion coefficients.

It is worth noting that the design, processing and installation of optical systems can achieve high precision, especially for high-priced lenses, so the distortion of thin prisms is very small and can usually be ignored. Generally, only radial distortion and tangential distortion are considered, and then only the first two-order distortion coefficients of each distortion can be considered. Even when the accuracy requirements are not too high or the focal length of the lens is long, only radial distortion can be considered.

4.1.4 Overview of Binocular Vision

The monocular vision of mobile robots can deduce 3D information from the 2D features of the image when the shape and properties of the object are known or obey certain assumptions, but in general, it is impossible to directly obtain the 3D environment information from a single image.

The binocular vision ranging method is a ranging method that imitates human beings to use binocular perception distance. Human eyes observe the scene of the objective three-dimensional world from two slightly different angles. Due to the projection of geometric optics, the images of object points at different distances from the observer on the retinas of the left and right eyes are not at the same position. This position difference on the retinas of the two eyes is called binocular parallax, which reflects the depth (or distance) of the objective scene. A binocular stereo camera is composed of two monocular cameras with a fixed positional relationship. First, two or more identical cameras are used to image the same scene from different positions to obtain a stereo image pair, and the corresponding image points are matched through various algorithms to calculate the parallax, and then the method based on triangulation is used to restore the distance. The difficulty of stereo vision ranging is how to choose reasonable matching features and matching criteria to ensure the accuracy of matching.

Compared with the monocular camera, the stereo camera model needs to add two matrices to correspond to the positional relationship between the two cameras. As shown in Fig. 4.2, R and T are the rotation matrix and the translation matrix, respectively.

$$R = \begin{bmatrix} 1 & \cos\theta & \sin\theta \\ \cos\theta & 1 & \sin\theta \\ \sin\theta & \cos\theta & 1 \end{bmatrix} \tag{4.9}$$

$$T = \left[T_x, T_y, T_z\right] \tag{4.10}$$

Depth recovery of various scene points can be achieved by computing disparity. Note that due to the discrete nature of digital images, the parallax value is an integer. In practical applications, some special algorithms can be used to make the parallax calculation precision to sub-pixel level. Therefore, for a given set of camera parameters, an effective way to improve the precision of scene point depth calculation is to increase the baseline distance B, that is, to increase the parallax corresponding to the scene point. However, this large-angle stereo method also brings some problems. The main problems are:

- As the baseline distance increases, the common visual range of the two cameras decreases.
- If the parallax value corresponding to the scene point increases, the range for searching the corresponding point increases, and the chance of ambiguity increases.

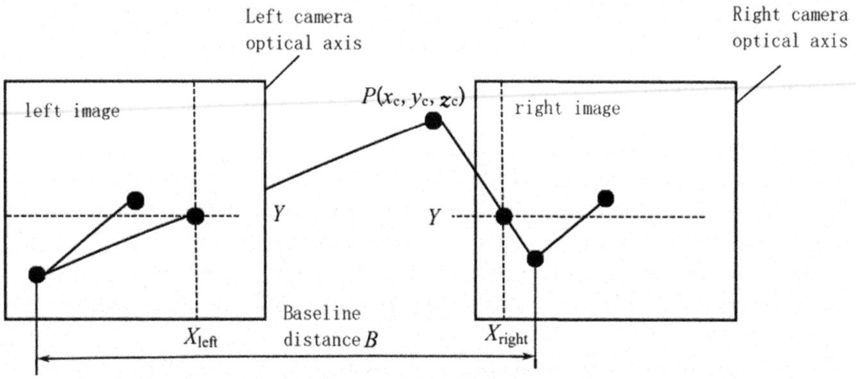

Fig. 4.2 Binocular stereo camera model

- The two images acquired by the two cameras are not exactly the same due to the distortion caused by perspective projection, which brings difficulties to the determination of conjugate pairs.

In practical applications, it is often encountered that the optical axes of the two cameras are not parallel, and the technology of adjusting their parallel coincidence is the calibration of the camera. When the two epipolar lines are not exactly in a straight line, that is, when the vertical parallax is not zero, for simplicity, many algorithms in the binocular stereo algorithm assume that the vertical parallax is zero.

4.2 Camera Calibration Method

Camera calibration is the basis of computer vision research and has important applications in 3D reconstruction and target tracking and positioning. The camera calibration method can be divided into offline calibration and online calibration according to the real-time calibration situation; according to the different calibration methods, it can be mainly classified into three types: traditional calibration method, self-calibration method and active vision-based calibration method.

The traditional calibration method refers to using a calibration block with a known structure and high precision as a spatial reference, establishing constraints on camera model parameters through the correspondence between spatial points and image points, and then obtaining these parameters through an optimization algorithm. The basic method is that, under a certain camera model, based on a known calibration reference object based on specific experimental conditions such as shape and size, after processing its image, a series of mathematical transformations and calculation methods are used to obtain the internal and external parameters of the camera model. The method is roughly divided into the basic method based on a single frame image and the stereo vision method based on the known correspondence of multiple frames.

Typical representatives of traditional methods include DLT method (Direct Linear Transformation), Tsai's method, and Weng's iterative method. The traditional calibration method, also known as strong calibration, is computationally complex and requires a calibration block, which is inconvenient, but it is suitable for any camera model and has high accuracy. The advantage of the traditional calibration method is that high precision can be obtained, but the calibration process is time-consuming and labor-intensive, and the calibration block cannot be used in many cases in practical applications, such as space robots, robots working in dangerous and harsh environments, etc. When the precision required in practical applications is very high and the parameters of the camera rarely change, the traditional calibration method should be the first choice.

The camera self-calibration method does not need to rely on any external special calibration objects or some control points whose three-dimensional information is known, but only uses the information of the corresponding points of the image to complete the calibration task directly through the image. It is this unique calibration idea that gives the camera self-calibration method great flexibility, and also enables computer vision technology to face a wider range of applications. The self-calibration method, also known as weak calibration, has low accuracy, belongs to nonlinear calibration, and is not robust, but only needs to establish the corresponding relationship between images, which is flexible and convenient, and does not require calibration blocks. It is well known that in many practical applications, the parameters of the camera often need to be changed, and the traditional camera calibration method becomes unsuitable in such cases due to the need to resort to special calibration objects. It is precisely because of its wide application and flexibility that the research on camera self-calibration technology has become one of the hotspots in the field of computer vision research in recent years.

Camera calibration based on active vision refers to a method of calibrating a camera under the condition that some motion information of the camera is known. Similar to the self-calibration methods, most of these methods only use the corresponding points of the image for calibration, and do not require high-precision calibration blocks. "Some motion information of the known camera" includes quantitative information and qualitative information: quantitative information, such as the camera translates a known amount in a certain direction in the platform coordinate system; qualitative information, such as the camera only moves in translation or only does rotational movement, etc. The active vision calibration method cannot be used in situations where the camera motion is unknown or uncontrollable, but it can usually be solved linearly and has strong robustness.

4.2.1 Offline Calibration Method

In most applications, real-time calibration is not required, and offline calibration is sufficient. In the indoor mobile robot navigation task, because the position of the stereo camera is fixed, and the parameters such as the focal length of the camera are also fixed, frequent calibration is not required, and the camera can be calibrated once without changing the camera. In view of the application environment, in order to obtain high-precision calibration and 3D measurement results, this chapter mainly studies the offline strong calibration method. So far, a variety of methods have been proposed for the strong calibration of cameras, which can be divided into three types according to different camera models: linear calibration method, nonlinear calibration method and two-step calibration method.

- Linear calibration

Direct Linear Transformation (DLT) was proposed by Abdel-Aziz and Karara in 1971. The linear method obtains the conversion parameters by solving the linear equation. The algorithm is fast, but the distortion of the camera lens is not considered, and the final result is very sensitive to noise. It is more suitable for the calibration of lenses with long focal length and small distortion. Due to its simplicity, direct linear transformation (DLT) is widely used in linear calibration methods.

- Nonlinear calibration

The more accurate the nonlinear model, the higher the computational cost. Since the nonlinear method takes into account the distortion of the camera lens, it uses a large number of unknowns and a wide range of nonlinear optimization, which makes the computational cost increase with the accuracy of the nonlinear model. Although the nonlinear optimization method has high precision, its algorithm is cumbersome and slow, and the iterative nature of the algorithm requires a good initial estimate. If the iterative process is not properly designed, especially under high distortion conditions, the optimization process may be unstable, resulting in unstable or even erroneous results, so its effectiveness is not high.

- Two-step calibration

Among the two-step calibration methods, Tsai's two-step calibration method is the most representative. This method only considers radial distortion, with moderate computational complexity and high accuracy. Weng proposed a nonlinear distortion model for CCD stereo vision, considering the sources of camera distortion, such as radial, centrifugal and thin prism distortion, and introduced a rotation matrix correction method, but it is difficult to achieve high precision by using matrix decomposition to find the initial values of internal and external parameters. In recent years, domestic scholars have also proposed two-step linear transformation methods. This two-step method only considers radial distortion and does not include nonlinear transformation, and can also achieve high accuracy. Zhang's two-step plane template method breaks away from the traditional method of sampling calibration images on a high-precision calibration stage. The calibration board can be manually placed at any position and in any attitude to calibrate the internal parameters of the camera.

– Principle of Tsai's two-step method

Tsai researches and summarizes the traditional calibration methods before 1987. On this basis, he proposes a practical two-step calibration algorithm for the camera model with radial distortion factor. For lenses with medium and long focal lengths, or high-priced lenses with small distortion rates, the Tsai two-step calibration method can achieve higher calibration and measurement precision. The algorithm is performed in two steps. The first step is to calculate the external parameters by establishing and solving the overdetermined linear equation system based on the image point coordinates with only radial distortion error; the second step, considering the distortion factor, use a three-variable optimization search algorithm to solve the nonlinear equation group to determine other parameters.

Assuming that the image coordinates (u_0, v_0) of the optical center have been obtained, in order to simulate the error in the installation process, an uncertainty factor S_X is introduced in the direction, and only the second-order radial distortion is considered for the distortion. Assume

$$\begin{cases} X_{di} = d_u(u_i - u_0) \\ Y_{di} = d_v(v_i - v_0) \end{cases} \tag{4.11}$$

Then there are:

$$\begin{cases} s_x^{-1}(1 + k_1 r^2)X_{di} = f\frac{r_{11}x_{wi}+r_{12}y_{wi}+r_{13}z_{wi}+t_x}{r_{31}x_{wi}+r_{32}y_{wi}+r_{33}z_{wi}+t_z} \\ (1 + k_1 r^2)Y_{di} = f\frac{r_{21}x_{wi}+r_{22}y_{wi}+r_{23}z_{wi}+t_y}{r_{31}x_{wi}+r_{32}y_{wi}+r_{33}z_{wi}+t_z} \end{cases} \tag{4.12}$$

that is:

$$\begin{aligned} &X_{di}(r_{21}x_{wi} + r_{22}y_{wi} + r_{23}y_{wi} + t_y) \\ &= s_x Y_{di}(r_{11}x_{wi} + r_{12}y_{wi} + r_{13}z_{wi} + t_x) \end{aligned} \tag{4.13}$$

a. Linear transformation to determine external parameters

Use more than 7 calibration points. According to the least square method, calculate the intermediate variable $t_y^{-1}s_x r_{11}, t_y^{-1}s_x r_{12}, t_y^{-1}s_x r_{13}, t_y^{-1}r_{21}, t_y^{-1}r_{22}, t_y^{-1}r_{23}, T_y^{-1}s_x t_x$ according to formula (4.14)

$$[Y_{di}x_{wi}\ Y_{di}y_{wi}\ Y_{di}z_{wi}\ Y_{di} - X_{di}x_{wi} - X_{di}y_{wi} - X_{di}z_{wi}]\begin{bmatrix} t_y^{-1}s_x r_{11} \\ t_y^{-1}s_x r_{12} \\ t_y^{-1}s_x r_{13} \\ t_y^{-1}s_x t_x \\ t_y^{-1}r_{21} \\ t_y^{-1}r_{22} \\ t_y^{-1}r_{23} \end{bmatrix} = X_{di} \tag{4.14}$$

Solve for external parameters $|t_y|$

Assume: $a_1 = t_y^{-1}s_x r_{11}, a_2 = t_y^{-1}s_x r_{12}, a_3 = t_y^{-1}s_x r_{13}, a_4 = t_y^{-1}s_x t_x, a_5 = t_y^{-1}r_{21}, a_6 = t_y^{-1}r_{22}, a_7 = t_y^{-1}r_{23}$, then:

$$|t_y| = (a_5^2 + a_6^2 + a_7^2)^{-1/2} \quad (4.15)$$

Determine the symbol of t_y

Use the image coordinates (u_i, v_i) and world coordinates (x_{wi}, y_{wi}, z_{wi}) of any feature point far from the center of the image for verification. That is: first assume $t_y > 0$, find out $r_{11}, r_{12}, r_{13}, r_{21}, r_{22}, r_{23}, t_x$, and $x = r_{11}x_{wi} + r_{12}y_{wi} + r_{13}z_{wi} + t_x$ and $y = r_{21}x_{wi} + r_{22}y_{wi} + r_{23}z_{wi} + t_y$, if X_{di} and x have the same symbol, Y_{di} and y have the same symbol, then t_y is positive, otherwise t_y is negative.

Define S_x

$$s_x = (a_1^2 + a_2^2 + a_3^2)^{1/2}|t_y| \quad (4.16)$$

Calculate R and t_x

$r_{11} = a_1t_y/s_x, r_{12} = a_2t_y/s_x, r_{13} = a_3t_y/s_x, r_{21} = a_5t_y, r_{22} = a_6t_y, r_{23} = a_7t_y,$ $t_x = a_4t_y/s_x, r_{31} = r_{12}r_{23} - r_{13}r_{22}, r_{32} = r_{13}r_{21} - r_{11}r_{23}, r_{33} = r_{11}r_{22} - r_{12}r_{21}$ ◦

b. Non-linear transformation to calculate internal parameters

Ignoring lens distortion, calculate rough values for f and t_z. (Assume $k_1 = 0$)

$$\begin{bmatrix} y_i & -Y_{di} \\ s_x x_i & -X_{di} \end{bmatrix} \begin{bmatrix} f \\ t_z \end{bmatrix} = \begin{bmatrix} w_i Y_{di} \\ w_i X_{di} \end{bmatrix} \quad (4.17)$$

Among, $x_i = r_{11}x_{wi} + r_{12}y_{wi} + r_{13}z_w + t_x$; $y_i = r_{21}x_{wi} + r_{22}y_{wi} + r_{23}z_w + t_y$; $w_i = r_{31}x_{wi} + r_{32}y_{wi} + r_{33}z_w$. Use the least squares method to find the rough values of f and t_z for the calibration points.

c. Calculate the exact f, t_z, k_1.

Using the f and t_z calculated above as the initial value (least square method), take the initial value of k_1 as 0.

$$\begin{cases} Y_{di}(1 + k_1r^2) = \frac{fy_i}{w_i + t_z} \\ s_x^{-1}X_{di}(1 + k_1r^2) = \frac{fx_i}{w_i + t_z} \end{cases} \quad (4.18)$$

Perform nonlinear optimization on the above formula to solve f, tz, and k1.

The optimization function is $\sum_{i=1}^{n}\left(\left(Y_{di}\left(1 + k_1r^2\right) - \frac{fy_i}{w_{i+t_z}}\right)^2 + \left(s_x^{-1}X_{di}\left(1 + k_1r^2\right) - \frac{fx_i}{w_i + t_z}\right)^2\right)$, which is the residual sum of squares of the equations. The above process is repeated for two camera calibrations.

– Zhang's two-step principle of plane template

Since the traditional calibration method requires a high-precision calibration stage, the calibration process is more complicated. Zhang proposed a method between the traditional calibration method and the self-calibration method, which avoids the high-precision calibration stage required by the traditional method. The operation is simple and the presicion is higher than that of the self-calibration method. First, the ideal perspective model is solved with the point near the center of the image, and the initial value is accurately estimated, and then the actual imaging model is solved with the full field of view calibration point. The calibration principle is as follows.

First, the mapping relationship between the target plane and the image plane is established, and the three-dimensional point coordinates on the target are marked as $\mathbf{M} = [x, y, z]^T$, the coordinates of the points on the image plane are marked as $\mathbf{m} = [u, v]^T$, The corresponding homogeneous coordinates are $\mathbf{M}' = [x, y, z, 1]'$, $\mathbf{m}' = [u, v, 1]^T$, and the relationship between the space point M and the image point m is:

$$s\mathbf{m}' = \mathbf{A}[\,\mathbf{R}\ \mathbf{t}\,]\mathbf{M}' \tag{4.19}$$

Among them, s is a non-zero scale factor, R, t are the rotation matrix and translation vector, which are the external parameters of the camera, and A is the internal parameters of the camera.

$$\mathbf{A} = \begin{bmatrix} \alpha\ c\ u_0 \\ 0\ \beta\ v_0 \\ 0\ 0\ 1 \end{bmatrix} \tag{4.20}$$

Assuming that the target plane is located on the x–y plane of the world coordinate system, that is, z = 0, it can be obtained from the above formula:

$$s\begin{bmatrix} u \\ v \\ 1 \end{bmatrix} = A\begin{bmatrix} \mathbf{r}_1\ \mathbf{r}_2\ \mathbf{r}_3\ \mathbf{t} \end{bmatrix}\begin{bmatrix} x \\ y \\ 0 \\ 1 \end{bmatrix} = A\begin{bmatrix} \mathbf{r}_1\ \mathbf{r}_2\ \mathbf{t} \end{bmatrix}\begin{bmatrix} x \\ y \\ 1 \end{bmatrix} \tag{4.21}$$

Let

$$\mathbf{H} = \mathbf{A}[\mathbf{r}_1, \mathbf{r}_2, \mathbf{t}] \tag{4.22}$$

The above formula can be written as:

$$s\begin{bmatrix} u \\ v \\ 1 \end{bmatrix} = \begin{bmatrix} h_{11}\ h_{12}\ h_{12} \\ h_{21}\ h_{22}\ h_{23} \\ h_{31}\ h_{32}\ h_{33} \end{bmatrix}\begin{bmatrix} x \\ y \\ 1 \end{bmatrix} \tag{4.23}$$

If the coordinates of multiple 3D points and their corresponding image coordinates are known, the H matrix can be solved with the following formula:

$$\mathbf{LH} = 0 \tag{4.24}$$

Among:

$$\mathbf{L} = \begin{bmatrix} x_1 & y_1 & 1 & 0 & 0 & 0 & -u_1x_1 & -u_1y_1 & -u_1 \\ 0 & 0 & 0 & x_1 & y_1 & 1 & -v_1x_1 & -v_1y_1 & -v_1 \\ & \cdots & & & \cdots & & & & \\ x_n & y_n & 1 & 0 & 0 & 0 & -u_nx_n & -u_ny_n & -u_n \\ 0 & 0 & 0 & x_n & y_n & 1 & -v_nx_n & -v_ny_n & -v_n \end{bmatrix} \tag{4.25}$$

$$\mathbf{H} = \begin{bmatrix} h_{11} & h_{12} & h_{13} & h_{21} & h_{22} & h_{23} & h_{31} & h_{32} & h_{33} \end{bmatrix} \tag{4.26}$$

The optimal solution of $\begin{bmatrix} h_{11} & h_{12} & h_{13} & h_{21} & h_{22} & h_{23} & h_{31} & h_{32} & h_{33} \end{bmatrix}$ can be solved using the least squares method. The value of each component of L is relatively large, and the coefficient matrix is actually an ill-conditioned matrix, which needs to be normalized and divided.

H can be written as $[\mathbf{h_1}\ \mathbf{h_2}\ \mathbf{h_3}]=\lambda\mathbf{A}[\mathbf{r}_1, \mathbf{r}_2, \mathbf{t}]$, $\mathbf{h_1}\ \mathbf{h_2}\ \mathbf{h_3}$ is the column vector of H, λ is any scalar, we can get:

$$\begin{cases} \mathbf{r}_1 = \lambda\mathbf{A}^{-1}\mathbf{h_1} \\ \mathbf{r}_2 = \lambda\mathbf{A}^{-1}\mathbf{h}_2 \end{cases} \tag{4.27}$$

From the orthogonality of the rotation matrix, we can get:

$$\begin{cases} \mathbf{h_1^T A^{-T} A^{-1} h_2^T} = 0 \\ \mathbf{h_1^T A^{-T} A^{-1} h_1^T} = \mathbf{h_2^T A^{-T} A^{-1} h_2^T} \end{cases} \tag{4.28}$$

Let:

$\mathbf{B} = \mathbf{A}^{-\mathbf{T}}\mathbf{A}^{-1} = \begin{bmatrix} B_{11} & B_{12} & B_{13} \\ B_{21} & B_{22} & B_{23} \\ B_{31} & B_{32} & B_{33} \end{bmatrix}$, then:

$$\mathbf{B} = \begin{bmatrix} \frac{1}{\alpha^2} & -\frac{c}{\alpha^2\beta} & \frac{cv_0-u_0\beta}{\alpha^2\beta} \\ -\frac{c}{\alpha^2\beta} & \frac{c^2}{\alpha^2\beta^2}+\frac{1}{\beta^2} & -\frac{c(cv_0-u_0\beta)}{\alpha^2\beta^2}-\frac{v_0}{\beta^2} \\ \frac{cv_0-u_0\beta}{\alpha^2\beta} & -\frac{c(cv_0-u_0\beta)}{\alpha^2\beta^2}-\frac{v_0}{\beta^2} & \frac{c(cv_0-u_0\beta)^2}{\alpha^2\beta^2}+\frac{v_0^2}{\beta^2}+1 \end{bmatrix} \tag{4.29}$$

Since B is a symmetric matrix, a vector $\mathbf{b} = \begin{bmatrix} B_{11} & B_{12} & B_{22} & B_{13} & B_{23} & B_{33} \end{bmatrix}^T$ can be defined, and let the i th column vector of matrix $\mathbf{H}$ be $\mathbf{h}_i = \begin{bmatrix} \mathbf{h}_{i1} & \mathbf{h}_{i2} & \mathbf{h}_{i3} \end{bmatrix}^T$, there are:

$$\mathbf{h}_i^T\mathbf{B}\mathbf{h}_j^T = \begin{bmatrix} \mathbf{h}_{i1} & \mathbf{h}_{i2} & \mathbf{h}_{i3} \end{bmatrix} \begin{bmatrix} B_{11} & B_{12} & B_{13} \\ B_{21} & B_{22} & B_{23} \\ B_{31} & B_{32} & B_{33} \end{bmatrix} \begin{bmatrix} \mathbf{h}_{i1} \\ \mathbf{h}_{i2} \\ \mathbf{h}_{i3} \end{bmatrix} = \mathbf{v}_{ij}^T\mathbf{b} \tag{4.30}$$

Among .$v_{ij} = \left[h_{i1}h_{j1}h_{i1}h_{j2} + h_{i2}h_{j1}h_{i2}h_{j2}h_{i3}h_{j1} + h_{i1}h_{j3}h_{i3}h_{j2}h_{i3}h_{j3}\right]^T$
The above formula can be rewritten as (4.31):

$$\mathbf{Vb} = 0 \tag{4.31}$$

Among $\mathbf{V} = \begin{bmatrix} \mathbf{v}_{12}^T \\ (\mathbf{v}_{11} - \mathbf{v}_{22})^T \end{bmatrix}$.

If n images are taken on the target plane, and these n equations are superimposed, if $n \geq 3$, then b is uniquely determined in the sense of a scale factor difference. When b is known, the elements of the A matrix can be solved:

$$\begin{aligned} u_0 &= (B_{12}B_{13} - B_{11}B_{23})/(B_{11}B_{22} - B_{12}^2) \\ \lambda &= B_{33} - [B_{13}^2 + v_0(B_{12}B_{13} - B_{11}B_{23})]/B_{11} \\ \alpha &= \sqrt{\lambda/B_{11}} \\ \beta &= \sqrt{\lambda B_{11}/(B_{11}B_{22} - B_{12}^2)} \\ c &= -B_{12}\alpha^2\beta/\lambda \\ u_0 &= cv_0/\alpha - B_{13}\alpha^2/\lambda \end{aligned} \tag{4.32}$$

A is known, $\mathbf{r}_1$, $\mathbf{r}_2$ can be solved from formula (4.27), $\mathbf{r}_3 = \mathbf{r}_1 \times \mathbf{r}_2$ can be obtained from the properties of the orthogonal matrix, and $\mathbf{t} = \lambda\mathbf{A}^{-1}\mathbf{h}_3$ can also be obtained from the above. So far, all the internal and external parameters of the camera have been obtained, and the parameters obtained in this way do not consider lens distortion, and the parameters obtained above are used as initial values for optimization search, and the optimal solution can be obtained.

4.2.2 Improved Node Extraction Algorithm

In Zhang's method, most of the nodes of the calibration board use the semi-automatic detection method, that is, the number and size of the checkerboard are artificially specified, and the software finds the nodes within the specified range. After many experiments, this method has the problem of inaccurate node search. According to this situation, the Harris node detector combined with the curve fitting method is used to improve the node extraction algorithm in the traditional algorithm, and realize the accurate positioning of nodes at sub-pixel level.

- Harris operator

The basic idea of Harris node extraction operator is similar to that of Moravec algorithm, but it has been greatly improved. It has high stability and reliability, and can accurately extract nodes under the conditions of image rotation, grayscale change and noise interference. The operation of the Harris operator is all based on the first-order differentiation of the image.

Let the image brightness function f (x, y), define a local autocorrelation function E (dx, dy) to describe the brightness change of the point (x, y) on the image after a small movement (dx, dy). The change of brightness is represented by the convolution of the square of the pixel brightness change value and the Gaussian function in the square neighborhood of the radius w around the (x, y) point:

$$E(dx, dy) = \sum_{i=x-w}^{x+w} \sum_{j=y-w}^{y+w} G(x-i, y-j)[f(i+dx, j+dy) - f(i, j)]^2 \tag{4.33}$$

where $G(x, y)$ is a two-dimensional Gaussian function. Expand the above formula by Taylor series:

$$\begin{aligned} E(dx, dy) = & \sum_{i=x-w}^{x+w} \sum_{j=y-w}^{y+w} G(x-i, y-j) \\ & \left[dx\frac{\partial f(i, j)}{\partial x} + dy\frac{\partial f(i, j)}{\partial y} + o\left(dx^2 + dy^2\right)\right]^2 \\ & \approx Adx^2 + 2Cdxdy + Bdy^2 \end{aligned} \tag{4.34}$$

Among,

$$A = G * \left(\frac{\partial f}{\partial x}\right)^2, B = G * \left(\frac{\partial f}{\partial y}\right)^2, \mathrm{C} = G * \left(\frac{\partial f}{\partial x}\frac{\partial f}{\partial y}\right) \tag{4.35}$$

Write the real quadratic form E (dx, dy) in matrix form:

$$E(dx, dy) = [dx, dy]M\begin{bmatrix} dx \\ dy \end{bmatrix} \tag{4.36}$$

Among, $M = \begin{bmatrix} A\,C \\ C\,B \end{bmatrix}$.

Then the matrix M describes the shape of the quadric surface z = E (dx, dy) at the origin. Let α, β be the eigenvalues of the matrix M, then α, β are proportional to the two principal curvatures, and are rotation invariants about M. If the two principal curvatures are very small, the surface z = E (dx, dy) is in the plane, and the movement of the detection window in any direction will not cause a large change in E, indicating that the gray levels in the detection window area are roughly the same. If one principal curvature is high and the other is low, the surface z = E (dx, dy) is ridge-like, and only movement in the direction of the ridge (the edge direction) causes a small change in E, indicating that the detected point is an edge point. If both principal curvatures are high, the surface z = E (dx, dy) is a downward spike, and the movement of the detection window in any direction will cause a rapid increase in E, indicating that the detected point is a node.

By the properties of the eigenvalues:

$$tr(M) = \alpha + \beta = A + B \tag{4.37}$$

$$\det(M) = \alpha\beta = AB - C^2 \tag{4.38}$$

The corner response function of the Harris operator is defined as:

$$CRF = \det(M) - k \cdot tr^2(M) \tag{4.39}$$

where k is a constant, Harris suggested to take 0.04. The node response value CRF is positive in the node region, negative in the edge region, and very small in the region where the grayscale remains unchanged. Because the node response value will increase at higher image contrast.

In the actual algorithm, the following strategies are adopted to further improve the performance of Harris operator:

- The disadvantage of Harris operator is that it is time-consuming. The reason is that 3 Gaussian smoothing is required to detect each point. If the points with lower gradient amplitudes are excluded, the efficiency can be greatly improved. In the algorithm, the image is first convolved with the Sobel operator, and the first-order differentials $\partial f / \partial x$ and $\partial f / \partial y$ in the directions x and y of each pixel are calculated. If the gradient amplitude is large ($(|\partial f / \partial x| + |\partial f / \partial y|)$ is greater than a certain threshold), $(\partial f / \partial x)^2$, $(\partial f / \partial y)^2$ and $(\partial^2 f / \partial x \partial y)$ are calculated, otherwise these three values are set to 0. Then, for the pixels with large gradient amplitude, convolve with a two-dimensional Gaussian function in its neighborhood window, and the size of the window is determined according to the standard deviation σ of the Gaussian function according to the 3σ criterion.
- For images that are greatly affected by noise, there will be multiple high responses near a node. Simply setting the threshold cannot completely eliminate false detections, but will miss some nodes with weak contrast. Nodes can be determined by the following method: After calculating the node response value CRF of each pixel, if the CRF of a certain pixel is the highest in its neighborhood, it is regarded as a node. The size of the neighborhood is set according to the specific image and needs. If the nodes in the detailed part of the image need to be detected, the neighborhood should be set smaller; if only obvious nodes need to be detected, the neighborhood should be enlarged. According to this method, the positioning performance of Harris operator can be improved.

It must be pointed out that the Harris operator can only be accurate to the pixel level. In order to further improve the accuracy, curve fitting is used to calibrate the edge of the board, and the node value is obtained through the intersection of the straight lines. Through this algorithm, the pixel accuracy is accurate to the sub-pixel level.

- Curve fitting method

According to the characteristics of the checkerboard calibration board, the fitting method is used to find the sub-pixel coordinates of the nodes. Each node is a common vertex of the four squares. Using the fitting method, the edge of the square can be fitted to a straight line, and each node is the intersection of the two straight lines.

Assuming a given data point (x_i, y_i) $(i = 0,1,\ldots,m)$, Φ is a function class composed of all polynomials whose degree does not exceed n $(n \leq m)$, now find a $p_n(x) = \sum_{k=0}^{n} a_k x^k \in \Phi$ such that:

$$I = \sum_{i=0}^{m} [P_n(x_i) - y_i]^2 = \sum_{i=0}^{m} \left(\sum_{k=0}^{n} a_k x_i^k - y_i \right)^2 = \min \quad (4.40)$$

When the fitting function is a polynomial, it is called fitting polynomial, and $p_n(x)$ satisfying Eq. (4.40) is called least square fitting polynomial. In particular, when n = 1, it is called linear fitting or straight line fitting.

Obviously, $I = \sum_{i=0}^{m} \left(\sum_{k=0}^{n} a_k x_i^k - y_i \right)^2$ is a multivariate function of $a_0, a_1, \ldots a_n$, so the above problem is the problem of finding the extreme value of $I = I(a_0, a_1, \ldots a_n)$. The necessary condition for finding the extremum by the multivariate function, we get:

$$\frac{\partial I}{\partial a_j} = 2 \sum_{i=0}^{m} \left(\sum_{k=0}^{m} a_k x_i^k - y_i \right) x_i^j = 0 \quad j = 0, 1, \ldots n \quad (4.41)$$

That is

$$\sum_{k=0}^{n} \left(\sum_{i=0}^{m} x_i^{j+k} \right) a_k = \sum_{i=0}^{m} x_i^j y_i, \quad j = 0, 1, \ldots n \quad (4.42)$$

Equation (4.42) is a system of linear equations with respect to a_0, a_1, ..., a_n, represented by a matrix as

$$\begin{bmatrix} m+1 & \sum_{i=0}^{m} x_i & \cdots & \sum_{i=0}^{m} x_i^n \\ \sum_{i=0}^{m} x_i & \sum_{i=0}^{m} x_i^2 & \cdots & \sum_{i=0}^{m} x_i^{n+1} \\ \vdots & \vdots & & \vdots \\ \sum_{i=0}^{m} x_i^n & \sum_{i=0}^{m} x_i^{n+1} & \cdots & \sum_{i=0}^{m} x_i^{2n} \end{bmatrix} \begin{bmatrix} a_0 \\ a_1 \\ \vdots \\ a_n \end{bmatrix} = \begin{bmatrix} \sum_{i=0}^{m} y_i \\ \sum_{i=0}^{m} x_i y_i \\ \vdots \\ \sum_{i=0}^{m} x_i^n y_i \end{bmatrix} \quad (4.43)$$

Equation (4.42) or Eq. (4.43) is called a normal equation system or a normal equation system.

It can be proved that the coefficient matrix of the system of Eqs. (4.42) is a symmetric positive definite matrix, so there is a unique solution. Solve a_k (k = 0,1,…,n) from Eq. (4.42), so that the polynomial

$$P_n(x) = \sum_{k=0}^{n} a_k x^k \tag{4.44}$$

It can be proved that $p_n(x)$ in Eq. (4.43) satisfies Eq. (4.41), that is, $p_n(x)$ is the required fitting polynomial. We call $\sum_{i=0}^{m} [p_n(x_i) - y_i]^2$ the squared error of the least squares fitting polynomial $p_n(x)$, and denote it as $\|r\|_2^2 = \sum_{i=0}^{m} [p_n(x_i) - y_i]^2$, which can be obtained from Eq. (4.44):

$$\|r\|_2^2 = \sum_{i=0}^{m} y_i^2 - \sum_{k=0}^{n} a_k \left(\sum_{i=0}^{m} x_i^k y_i \right) \tag{4.45}$$

The general method to polynomial fitting can be summarized in the following steps:

- Draw a rough graph of the function from the known data—a scatter plot, and determine the degree n of the fitting polynomial;
- List tables to calculate $\sum_{i=0}^{m} x_i^j (j = 0, 1, \ldots, 2n)$ and $\sum_{i=0}^{m} x_i^j y_i (j = 0, 1, \ldots, 2n)$;
- Write out the normal equations and find $a_0, a_1, \ldots a_n$;
- Write the fitting polynomial $p_n(x) = \sum_{k=0}^{n} a_k x^k$.

4.2.3 Experimental Results

- Camera calibration

Use the improved calibration algorithm proposed above to calibrate the stereo vision system equipped on the robot platform. The stereo camera is shown in Fig. 4.3, which is a split gigabit camera. The external parameters are: the baseline length is 7 cm, the two cameras are installed in parallel; the focal length of the camera lens is 8 mm.

The plane calibration board is a checkerboard, as shown in Fig. 4.4. Then fix the binocular stereo camera, take pictures of the calibration board from multiple angles, and use the obtained pictures to calibrate the stereo camera.

First, the traditional calibration method is used to extract the nodes, and the effect diagram is shown in Fig. 4.5, and then the algorithm we propose is used to extract the nodes, and the result diagram is shown in Fig. 4.6. Since the grid image is too small, the improvement result is not obvious, so the image is enlarged by 10 times, as

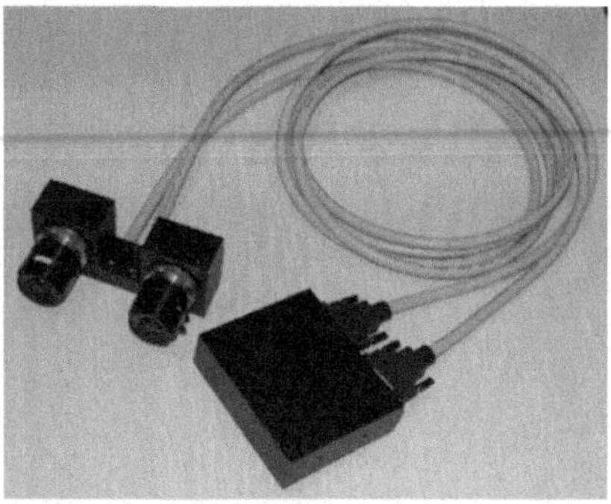

Fig. 4.3 Stereo camera

Fig. 4.4 Flat calibration plate

shown in Figs. 4.7 and 4.8. It can be seen from the figure that the improved algorithm extracting nodes can accurately find the nodes.

One of the main methods to test the accuracy of camera calibration is to calculate the pixel back projection error. The back projection pixel error using the traditional node extraction algorithm (as shown in Fig. 4.9) and the back projection pixel error using the improved node extraction algorithm (as shown in Fig. 4.10) are calculated respectively. It can be seen from the figure that the precision of the traditional algorithm is more than one pixel, while the improved node extraction algorithm can be accurate to 0.3 pixels.

- Ranging accuracy test

The ranging accuracy of a camera depends on many factors: such as the focal length of the lens, the accuracy of camera calibration, the distance to the object being measured, and so on. Different cameras need to be selected according to different uses. For indoor mobile robot navigation, the lens we choose is a 9 mm focal length lens with a baseline length of 7 cm, as shown in Fig. 4.11.

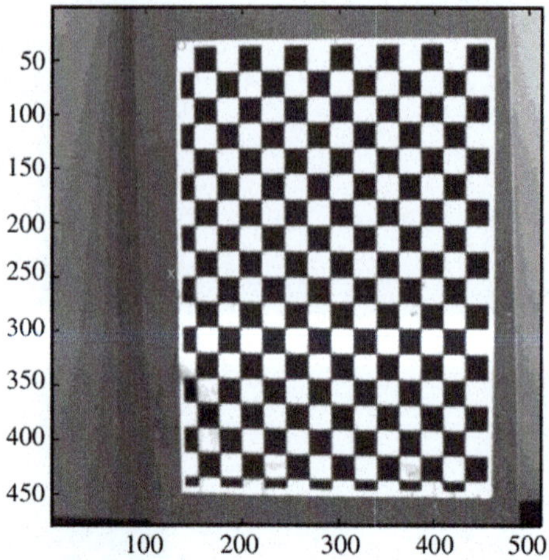

Fig. 4.5 Traditional node extraction renderings

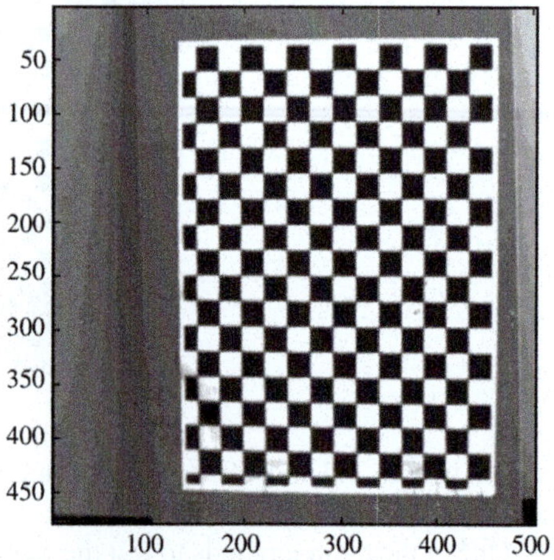

Fig. 4.6 Improved node extraction renderings

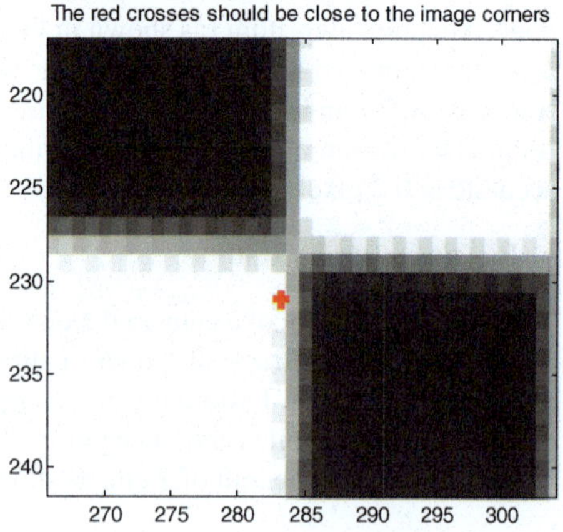

Fig. 4.7 Traditional node extraction renderings(10 × magnification)

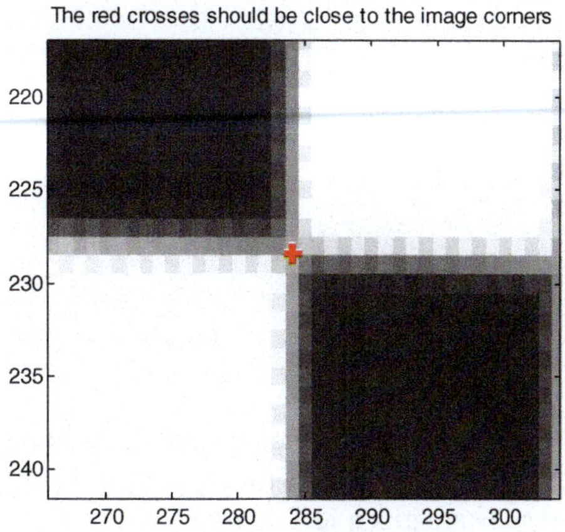

Fig. 4.8 Improved node extraction renderings(10 × magnification)

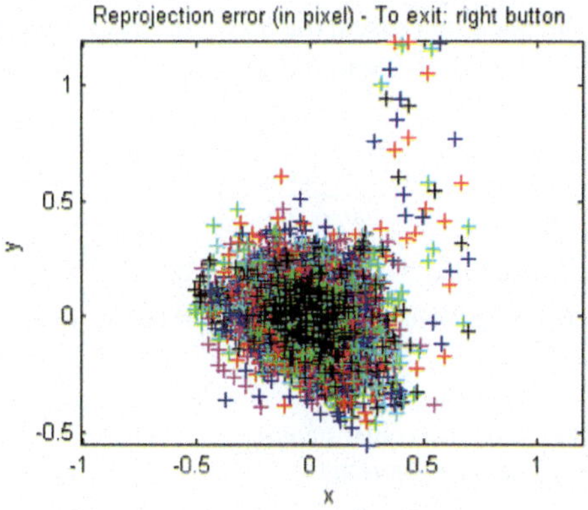

Fig. 4.9 Traditional backprojection pixel error

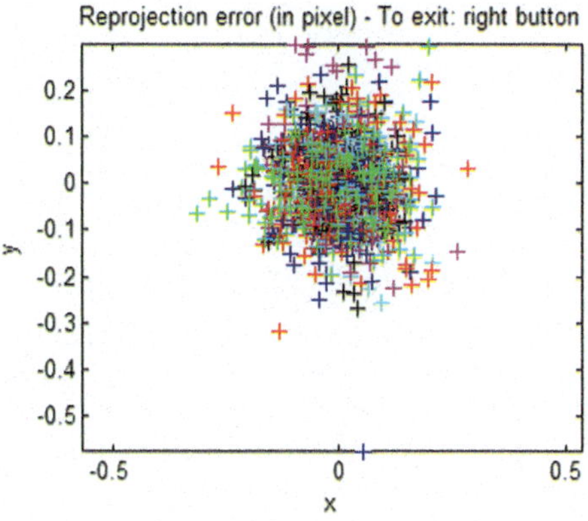

Fig. 4.10 Improved method back projection pixel error

Fig. 4.11 Binocular stereo camera

The test environment is that the vision is facing the target, and then the distance is gradually changed to test the precision of the binocular vision system. The test results are given in Table 4.1.

Figure 4.12 shows the error test results of the camera. It can be seen that as the distance increases, the error increases, and the distance range with the smallest error is between 1.2 and 3.2 m.

Table 4.1 Binocular stereo vision ranging experiment (mm)

Time	1	2	3	4	5	6	7	8	9	10	11
Actual value	1200	1500	1800	2100	2400	2700	3000	3300	3600	3900	4200
Measured value	1175	1470	1860	2140	2370	2650	3080	3220	3470	3750	4080

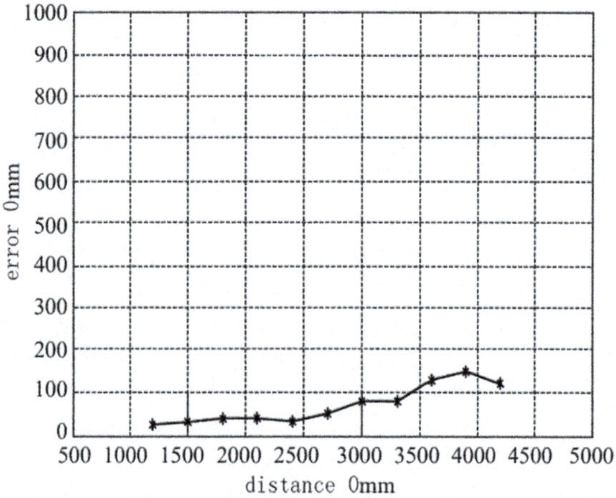

Fig. 4.12 Binocular stereo camera error test results

4.3 Design and Identification of Road Signs

In order to enable the mobile robot to achieve self-positioning relatively stably. The design of several road signs and the corresponding identification methods are proposed below.

4.3.1 *Border Design and Identification*

The design of artificial road signs for indoor mobile robot navigation mainly considers three requirements: reliability, real-time and aesthetics. Reliability requires that the robot can effectively and reliably detect and recognize road signs within the current field of view, and accurately calculate the robot's pose according to the road signs; real-time performance require that the road sign detection speed is fast, and can quickly calculate the pose according to the road sign; aesthetics is an aspect that is easy to be ignored. There is no need to overemphasize its importance in experimental research, but it is a crucial issue for commercial robot products.

For the above reasons, the road signs we designed should be easy to distinguish in the environment, and the recognition algorithm should be robust and fast. Therefore, the designed road sign consists of two parts, the first part is a red border, and the second part is an expandable road sign pattern, as shown in Fig. 4.13.

In the road sign experiment environment, the stereo camera may observe the road sign from any angle, so the feature quantity of the road sign identification is required to be immune to rotation and scaling, and to have projective invariance. We use a combined recognition algorithm to identify road signs, including chromaticity, rectangularity, and intersection ratio invariants.

(1) Chromaticity

The first parameter that can stably express color information is chromaticity information. In order to make chromaticity information unaffected by light, we first convert the obtained image from RGB into HSV space. The HSV space hexagonal pyramid is shown in Fig. 4.14.

Fig. 4.13 A road sign

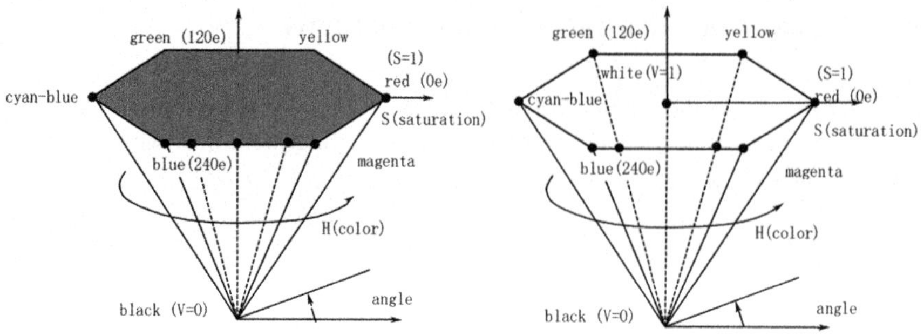

Fig. 4.14 HSV space hexagonal pyramid

Figure 4.14 shows the HSV color model. The lightness V gradually increases from 0 at the apex of the pyramid to the top surface along the axis and the maximum value of 1 is taken, the color saturation S is determined by the distance from the point on the pyramid to the central axis, and the color H is expressed as the angle between it and red. Red is placed at 00 in the figure. The color saturation value ranges from 0 on the axis to 1 on the outer edge. Only the fully saturated primary color and its complementary color have S = 1, At S = 0, the color H is undefined, and the corresponding color is a certain level of gray. Along the central axis, gray changes from light to dark, forming different layers.

The transformation formula from the RGB-HSV space is shown in Eqs. (4.46) to (4.48).

$$H = \begin{cases} \left(6 + \dfrac{G - B}{MAX - MIN}\right) \times 60^\circ, if \text{ R} = \text{MAX} \\ \left(2 + \dfrac{B - R}{MAX - MIN}\right) \times 60^\circ, if \text{ G} = \text{MAX} \\ \left(4 + \dfrac{R - G}{MAX - MIN}\right) \times 60^\circ, if \text{ B} = \text{MAX} \end{cases} \tag{4.46}$$

$$S = \frac{MAX - MIN}{MAX} \tag{4.47}$$

$$V = MAX \tag{4.48}$$

Because HSV space is very sensitive to color transformation, and has strong anti-interference to changes in illumination, so the image is transformed into HSV space, and a certain threshold is set to segment the image.

- Rectangularity

Rectangularity is described by the ratio of the area of the object to the area of its smallest surrounding rectangle, which reflects the fullness of the object to its exterior rectangle. Because the border we use is a rectangle, the introduction of rectangularity can quickly locate the road sign. The formula for calculating the squareness is:

$$R = A/A_{mer} \tag{4.49}$$

Among them, A is the area of the rectangle enclosed by the peripheral frame, and A_{mer} is the area of the minimum exterior rectangle.

- Intersection ratio invariants

The intersection ratio invariant is the most basic invariant in projective geometry. At four collinear points, as shown in Fig. 4.15, the definition of the intersection ratio invariant is shown in Eq. (4.50). Cross-ratio invariants in road signs is shown in Fig. 4.16.

$$R = (P_1P_2, P_3P_4) = \frac{(P_1P_2P_3)}{(P_1P_2P_4)} = \frac{P_1P_3 \cdot P_2P_4}{P_2P_3 \cdot P_1P_4} \tag{4.50}$$

To sum up, the identification of road signs adopts a combined detection algorithm, and the block diagram of the detection algorithm is shown in Fig. 4.17.

Part of the results of detecting road signs using the combination algorithm proposed above are given in Fig. 4.18.

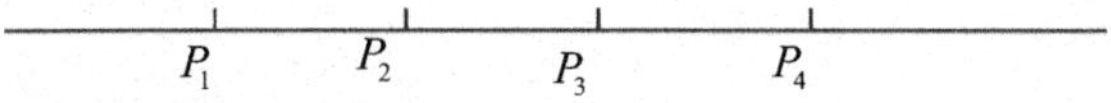

Fig. 4.15 Cross ratio invariant

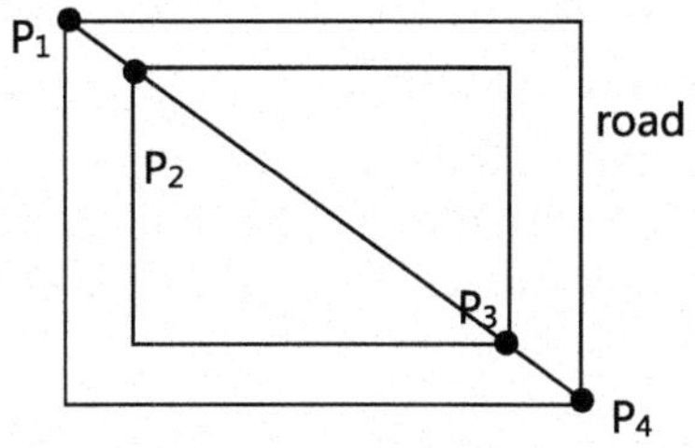

Fig. 4.16 Cross-ratio invariants in road signs

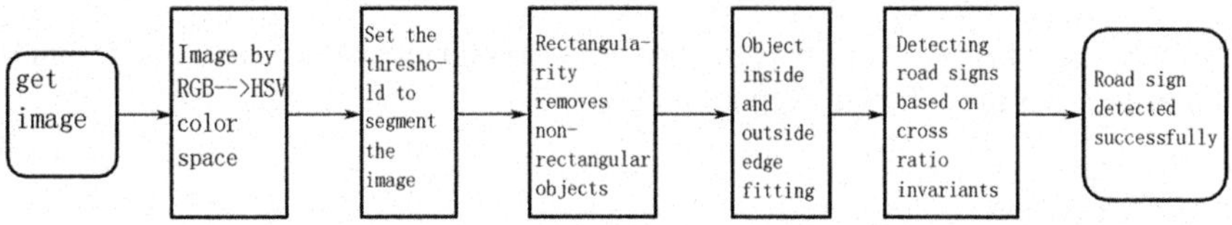

Fig. 4.17 Road sign detection algorithm block diagram

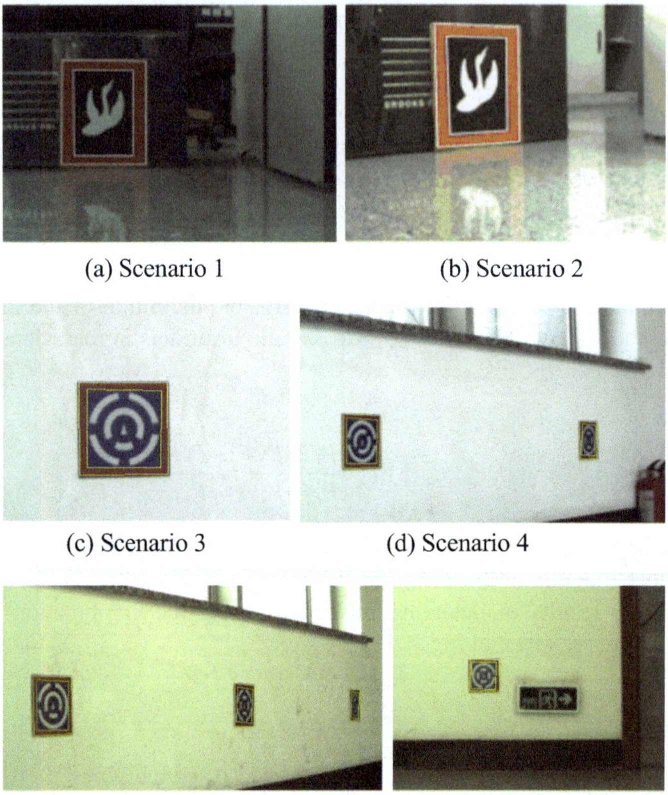

(a) Scenario 1 (b) Scenario 2

(c) Scenario 3 (d) Scenario 4

(e) Scenario 5 (f) Scenario 6

Fig. 4.18 Road sign detection results

According to the test results, it can be seen that for the various road signs we designed, whether in the case of sufficient light or in the case of dim light, the road signs can be detected relatively successfully, and the recognition success rate is more than 85%.

4.3.2 Pattern Design and Recognition

The second component of the road sign is an expandable pattern, and the design of the pattern has the following requirements:

- Strong scalability;
- Easy to detect;
- The probability of repetition with the surrounding environment is extremely small;
- The impact on the surrounding environment is small.

Based on the above requirements, a total of three schemes of road sign patterns have been designed in the experiment. The three artificial road signs designed are all flat artificial road signs, similar to the decorative paintings hanging on the wall, with good visual effects. The first scheme is an abstract animal pattern, as shown in Fig. 4.19, which can be extended to dozens of patterns; the second scheme is a two-digit Arabic numeral, with a total of 100 possibilities, as shown in Fig. 4.20; The third scheme is a circular road sign pattern, which can expand a total of 256 possibilities, as shown in Fig. 4.21.

According to the characteristics of the road sign pattern, we select the appropriate feature vector corresponding to each pattern, and then train through the support vector machine learning algorithm (SVM) to obtain the object sample library. In practical applications, it is only necessary to compare the actual captured image with all the samples in the sample library, and find the most suitable sample according to the minimum variance. The chosen feature vector is required to be immune to rotation, scaling, and translation of the image. Several types of image global feature vectors selected in the design are: image invariant moment, normalized moment of inertia (NMI), and multi-dimensional histogram.

- Invariant moment

The moment feature mainly characterizes the geometric feature of the image area, also known as the geometric moment, because it has the invariant features of rotation,

Fig. 4.19 Abstract animal graphic road sign scheme

Fig. 4.20 Digital road sign scheme

Fig. 4.21 Ring road sign scheme

translation, scale and other characteristics, so it is also called the invariant moment. The Hu invariant moment is a very commonly used image invariant moment. Hu moment is a nonlinear combination of regular moments, and the central moment of order p + q is:

$$\mu_{pq} = \sum_x \sum_y (x - \overline{x})^p (y - \overline{y})^q F(x, y) \tag{4.51}$$

where F(x, y) represents a two-dimensional image, and:

$$\overline{x} = m_{10}/m_{00}, \overline{y} = m_{01}/m_{00} \tag{4.52}$$

$$m_{pq} = \sum_x \sum_y x^p y^q F(x, y) \tag{4.53}$$

The normalized central moment of order p + q is:

$$\eta_{pq} = \mu_{pq}/\mu_{00}^r \tag{4.54}$$

In formula (4.64), r = 1 + (p + q)/2, p, q = 1, 2, 3.... Seven invariant moment groups $\Phi_1, \Phi_2, \ldots \Phi_7$ can be generated using the second-order and third-order normalized central moments. Hu invariant moments do not have affine invariance, and their higher-order invariant moments are sensitive to noise.

- Normalized Moment of Inertia (NMI)

The normalized moment of inertia has good translation, rotation and scaling invariance. Assuming that the center of gravity of the image grayscale is (cx, cy), the moment of inertia of the image around the centroid is recorded as $J_{(cx, cy)}$:

$$J_{(cx,cy)} = \sum_{x=1}^{M} \sum_{y=1}^{N} [(x - cx)^2 + (y - cy)^2] f(x, y) \tag{4.55}$$

Among them, $f(x, y)$ represents a two-dimensional image, M and N are the width and height of the image, respectively.

According to the definition of the centroid of the image and the moment of inertia, the NMI of the image around the centroid can be given as:

$$NMI = \frac{\sqrt{J_{(cx,cy)}}}{m} = \frac{\sqrt{\sum_{x=1}^{M}\sum_{y=1}^{N}[(x-cx)^2+(y-cy)^2]f(x, y)}}{\sum_{x=1}^{M}\sum_{y=1}^{N}f(x, y)} \tag{4.56}$$

where $\sum_{x=1}^{M}\sum_{y=1}^{N}f(x, y)$ is the image quality, representing the sum of all grayscale values in the image.

- Multidimensional histogram

Histogram is an important method to characterize various global features. When the histogram is used as the feature of pattern recognition, the higher the dimension of the histogram, the more effectively the image can generally be described, but it will lead to a rapid increase in the amount of computation. Assuming a 16-dimensional histogram, the quantization level of each dimension is 15, then the histogram contains a total of 15^{16} components. But in fact, the value of most components is 0. Therefore, the experiment adopts a histogram compression algorithm similar to hash to reduce the time complexity, that is, the so-called hybrid compressed histogram. Multidimensional histograms can be efficiently represented using one-dimensional compressed mixed histograms. Gradient direction, gradient magnitude, and color components are often used as image description operators for blended histograms. Composing different image description operators to form a compressed mixed histogram can effectively represent the information such as the structure and color of the image.

- Geometric template components

Corresponding to the third road sign scheme, the experiment uses a geometric template component as the feature component. The principle is shown in Fig. 4.22. The template component value includes 8 components, which are P_1–P_8 respectively, and the component value takes the gray level statistical value of a certain neighborhood around the point. The feature points can be calculated by the intersection ratio invariance theorem, such as P_5: on the premise that the frame is detected, Pa, Pc and the centroid O of the road sign are known, and three of the four points are known, then P_5 can be obtained according to the invariance theorem of intersection ratio:

$$R = \frac{P_a O \cdot P_5 P_7}{P_5 O \cdot P_a P_7} \tag{4.57}$$

By analogy, the positions of the remaining seven feature points can be obtained. According to the invariance theorem of intersection ratio, the template component is immune to rotation, translation and scaling, which is suitable for robot navigation.

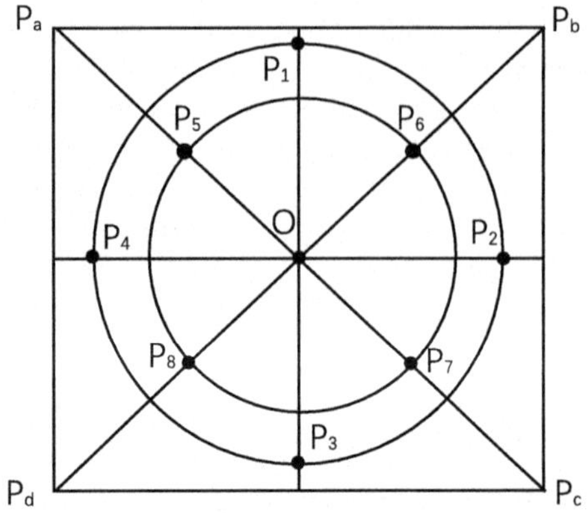

Fig. 4.22 Road sign geometric template components

Using the feature vector given above, SVM is used to train the template image to generate a standard sample library. In practical applications, the acquired image is extracted with features, and then compared with the samples in the standard library to find the most suitable target. After many experiments, the recognition rate of the first road sign is above 75%, the recognition rate of the second road sign is above 80%, and the recognition rate of the third road sign is above 86%.

4.4 Positioning System Based on Road Signs

The principle of the positioning system based on road signs is to use the stereo vision system to measure the three-dimensional coordinates of the road sign in the robot coordinate system. Assuming that the global coordinates of the road sign are known, Then the global coordinates of the robot in the global coordinate system can be calculated according to the local coordinates of the road sign in the robot coordinate system and the global coordinates in the global coordinate system.

4.4.1 Single Landmark Positioning System

As shown in Fig. 4.23, (X_W, O_W, Y_W) is the global coordinate system, (X_R, O_R, Y_R) is the local coordinate system of the robot, L is the top view of the road sign, $(x_{LW}, y_{LW}, \theta_{LW})$ is the coordinates of the center point of the road sign in the global coordinate system, θ_W is the angle between the direction X_L of the road sign and the direction X_W of the global coordinate system, $(x_{LR}, y_{LR}, \theta_{LR})$ is the local coordinate of the center point of the road sign in the visual measurement, then the coordinates of the robot body in the global coordinate system are:

Fig. 4.23 Single signposti

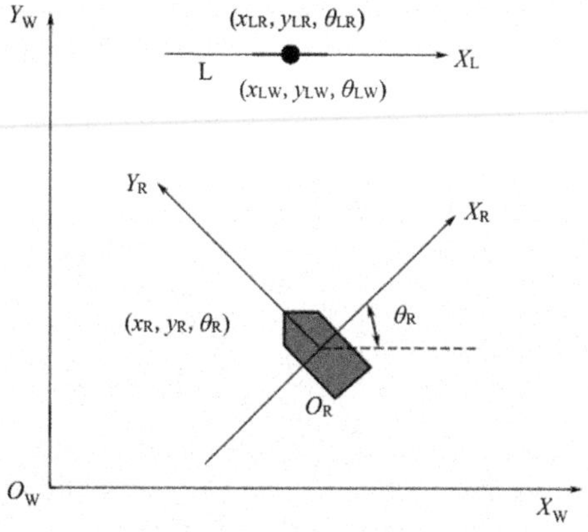

$$\begin{bmatrix} x_R \\ y_R \\ \theta_R \end{bmatrix} = \begin{bmatrix} x_{LW} \\ y_{LW} \\ \theta_{LW} \end{bmatrix} - \begin{bmatrix} \cos\theta_R & -sin\theta_R & 0 \\ \sin\theta_R & \cos\theta_R & 0 \\ 0 & 0 & 1 \end{bmatrix} \begin{bmatrix} x_{LR} \\ y_{LR} \\ \theta_{LR} \end{bmatrix} \tag{4.58}$$

4.4.2 Multiple Landmark Positioning System

There are two landmarks in the field of vision, Landmark1 and Landmark2, respectively, and the robot position expression formula is derived. The global coordinates and local coordinates of the center point of road sign 1 are: $\left(x_{LW}^1, Y_{LW}^1, \theta_{LW}^1\right)$, $\left(x_{LR}^1, Y_{LR}^1, \theta_{LR}^1\right)$; the global coordinates and local coordinates of the center point of road sign 2 are: $\left(x_{LW}^2, Y_{LW}^2, \theta_{LW}^2\right)$, $\left(x_{LR}^2, Y_{LR}^2, \theta_{LR}^2\right)$. The line segment N in Fig. 4.24 is the line connecting the left vertices of the two landmarks.

The position where the global positioning of the robot is derived is:

Fig. 4.24 Dual signposting

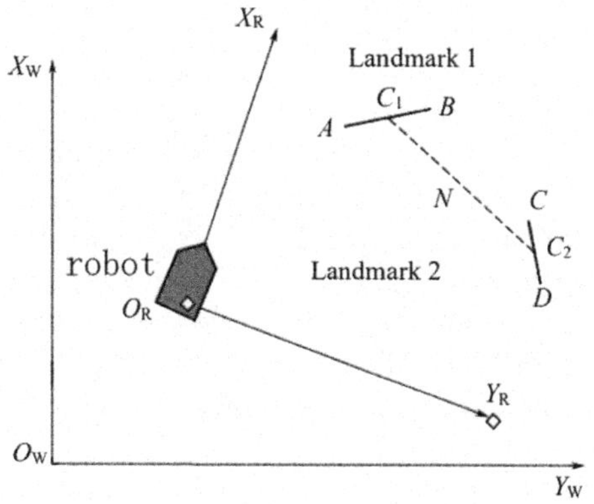

$$\theta_{NW} = \arctan\left(\frac{y_{LW}^2 - y_{LW}^1}{x_{LW}^2 - x_{LW}^1}\right) \tag{4.59}$$

$$\theta_{NR} = \arctan\left(\frac{y_{LR}^2 - y_{LR}^1}{x_{LR}^2 - x_{LR}^1}\right) \tag{4.60}$$

$$\begin{bmatrix} x_R \\ y_R \\ \theta_R \end{bmatrix} = \begin{bmatrix} x_{LW}^1 \\ y_{LW}^1 \\ \theta_{NW} \end{bmatrix} - \begin{bmatrix} \cos\theta_N & -\sin\theta_N & 0 \\ \sin\theta_N & \cos\theta_N & 0 \\ 0 & 0 & 1 \end{bmatrix} \begin{bmatrix} x_{LR}^1 \\ y_{LR}^1 \\ \theta_{NR} \end{bmatrix} \tag{4.61}$$

Among them, θ_{NW} is the angle between the line C_1C_2 connecting the midpoints of the two road signs and the x-axis of the global coordinate system, and θ_{NR} is the angle between C_1C_2 and the x-axis of the robot coordinate system.

When there are more than 2 road signs in the field of vision, each two road signs are a group, and the positions of $\frac{M!}{M}$ robots can be obtained, and then the final position of the robot can be obtained by the minimum variance method.

4.4.3 Error Analysis

Since the perception of the target by the stereo vision system is affected by environmental noise and illumination, and considering factors such as camera distortion and resolution, there is a certain uncertainty in the observation of the target by the stereo vision system.

The observation experiment of stereo vision proves that its observation characteristic is based on Gaussian distribution. A standard Gaussian distribution function is:

$$\mathrm{p(X)} = \frac{1}{2\pi\sqrt{|C|}} exp\left[-\frac{1}{2}(X - X')^T C(X - X')\right] \tag{4.62}$$

$$C = \begin{bmatrix} \sigma_x^2 & \rho_{\sigma_x\sigma_y} \\ \rho_{\sigma_x\sigma_y} & \sigma_y^2 \end{bmatrix} \tag{4.63}$$

Among them, X is the two-dimensional coordinate value $(x, y)^T$ of the position of the observation target, X is the mathematical expectation of the position, and C is the covariance matrix. The observation model of the robot based on the known position of the road sign in the local coordinate system is shown in Fig. 4.25, where σ_{max} and σ_{min} are the standard deviations on the main and short axes of the coordinate system. At this time, the correlation coefficient $\rho = 0$ of the two variances, the covariance matrix in the local coordinate system is:

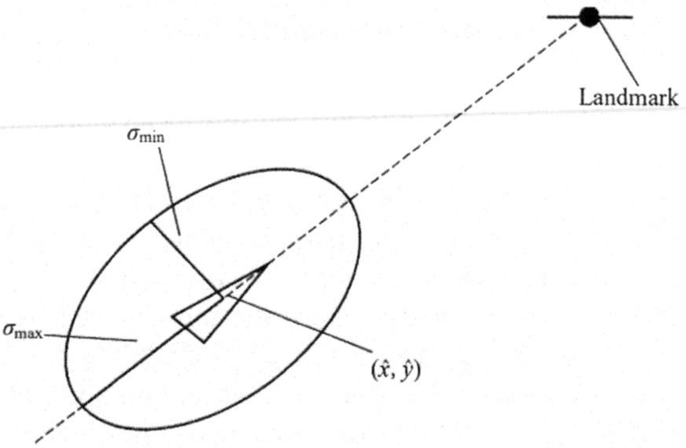

Fig. 4.25 Stereo vision based on road sign observation model

$$C_L = \begin{bmatrix} \rho_{\max}^2 & \\ & \rho_{\min}^2 \end{bmatrix} \tag{4.64}$$

Using multiple road signs to locate the robot can effectively reduce the error and improve the accuracy of the robot positioning. The principle is that after the observation covariance matrix of each landmark is obtained, all the covariance matrices can be fused, which can effectively reduce the range of errors.

$$C' = C_1 - C_2[C_1 + C_2]^{-1}C_1 \tag{4.65}$$

The above formula is the formula for covariance fusion calculation for dual-target stereo vision positioning. If more landmarks are successfully detected, all landmarks can be divided into two groups for fusion, and finally the covariance fusion results of all landmarks are obtained.

As can be seen from the Fig. 4.26, the error range of multiple road signs is smaller than that of single road sign self-positioning. If the field of view allows, theoretically, the more road signs are seen, the smaller the error of the robot's self-positioning.

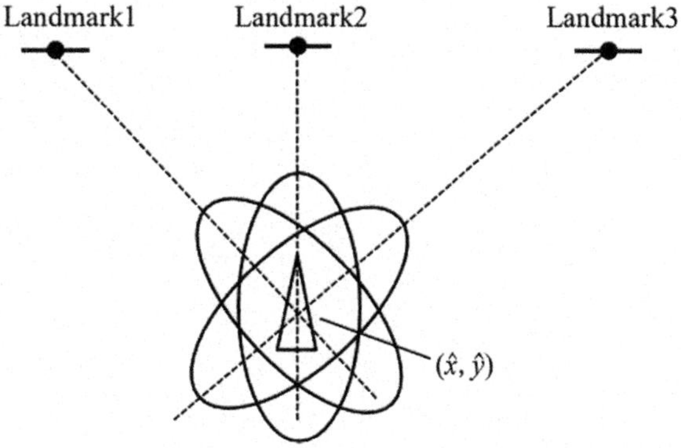

Fig. 4.26 Schematic diagram of multi-signpost data fusion

4.4.4 Experimental Verification

Experiments are used to test the self-localization precision of mobile robot based on multi-landmark stereo vision. The experimental site is a square site with a side length of 4 m, and road signs are placed inside the site, as shown in Fig. 4.27. The blue line segment in the figure is the top view where the road sign is placed. The preset route of the road sign is shown in the red "*" in the figure. It is considered that two working modes of the robot are set, one is self-positioning based on a single road sign, and the other is self-positioning based on multiple road signs. The black rhombus curve in the figure is the positioning position based on a single landmark, and the curve formed by the blue circle points is the positioning position of the robot based on multiple landmarks. It can be seen from the experimental results that the positioning precision of multiple landmarks is significantly better than that of single landmarks.

According to the experimental data, it can be seen that the maximum error of positioning with a single road sign can reach 500 mm, and the maximum error of the road sign when positioning with multiple road signs is reduced to less than 100 mm, and the error does not change drastically and does not accumulate, which also proves that the multi-road sign positioning method can effectively improve the self-positioning precision of the robot.

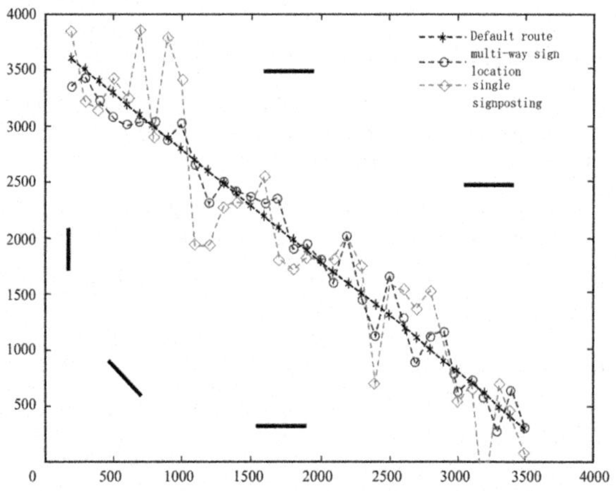

Fig. 4.27 Road sign positioning accuracy experiment

4.5 Analysis of Mobile Robot Positioning

4.5.1 Monte Carlo Localization Algorithm

Mobile robot positioning can be regarded as a Bayesian evaluation problem, that is, given input data, observation data, motion and perception models, and using prediction/update steps to estimate the robot's implicit pose state reliability optimization problem at the current moment. A typical evaluation state is generally $s = (x, y, \theta)$. where (x, y) represents the position of the robot in the Cartesian coordinate system; θ indicates the course angle of the robot. The input data u usually comes from the internal sensor odometer; the observation data z comes from external sensors such as lidar, camera, etc., and the motion model $p(s_t|s_{t-1}, u_{t-1})$ represents the initial state of the system at time t as s_{t-1}, and the input u_{t-1} is the probability of reaching the state S_t.

Monte Carlo localization, as a probabilistic localization method based on the principle of Bayesian filtering, is also realized by recursively estimating the probability distribution of the pose state space from the sensory information, but the probability distribution is described in the form of weighted sampling. The conventional Monte Carlo localization algorithm consists of three recursive steps in terms of implementation form, namely, prediction update (Prediction step), perception update (also known as Importance Sampling) and importance resampling process (Resampling).

For the conventional Monte Carlo algorithm, since only the motion model $p(X_k|X_{k-1})$ is used as the importance function, when there are some unmodeled robot motions, such as collision or kidnapping problems, the Monte Carlo method implemented with a small number of samples is difficult to solve. For the above problems, some of the solutions use the method of adaptive sampling number, and some introduce additional random uniformly distributed sampling into the proposal distribution. Although the above problems can be alleviated to a certain extent, the randomness of sampling selection will increase the unpredictability of the positioning process.

Considering the entire positioning process, since the sampling set after prediction update is uniformly distributed, the weight (unnormalized) of the sampling distribution updated by the perceptual information determines the matching degree between the sampling set and the current observation information. If a metric is used to test the matching degree, the resampling process can be introduced in a timely manner, and only using the motion model as the importance function cannot solve various positioning problems, and the re-sampling from the perceptual information is also introduced. In the recursive positioning process, the changes of the sampling distribution weights before and after the perceptual update are as follows.

First, when the sampling distribution matches the perceptual information well, the weights of most of the sampling sets after the perceptual update are still high (unnormalized), the distribution is relatively uniform, and these samples are still concentrated near the real pose of the robot, and the positioning error becomes smaller and smaller after the importance resampling, which is the process of pose tracking.

Secondly, when the sampling distribution does not completely match the perceptual information, there are two situations: first, considering the existence of various interferences, there may be a small number of high-weight samples, and the sampling distribution shows over-convergence; the other is to make the weight distribution of all samples more uniform, but the weight is small, these two cases correspond to the initial positioning process or the kidnapping robot problem, indicating that the current sampling distribution is no longer a good estimate of the robot pose distribution.

Aiming at the above analysis, an extended Monte Carlo localization algorithm is implemented, which uses the updated sampling distribution information as the judgment basis to perform resampling in a timely manner to save computing resources and improve the localization efficiency. In addition to the two recursive procedures of the conventional Monte Carlo positioning algorithm, two additional checking procedures, the over-convergence checking procedure and the uniformity checking procedure, are introduced in the algorithm. These two processes are used to judge how well the samples from the motion model match the perceptual information to introduce different resampling methods.

- Over-convergence test process

This process uses the information entropy and the effective number of samples to test the over-convergence of the sampling distribution for the normalized sampling weights. For example, when the initial sampling update is performed or the robot is kidnapped, the similarity of the environment model and the uncertainty of the perception information are considered. Since the sampling distribution does not completely match the perceptual information, the sampling weight distribution will inevitably be over-converged (a small number of sampling high weights, mass sampling low weights). When the effective sampling number is less than the given threshold, the sampling distribution is over-converged; otherwise, the over-convergence is determined according to the relative change of information entropy. Over-convergence then performs the process of resampling from perceptual information and resampling from importance, respectively.

- Uniformity inspection process

If the sampling distribution does not appear to converge, the matching of the sampling distribution and the perceptual information is checked according to the sum of the weights of the unnormalized sampling distribution. When the sum of the weights is greater than the given threshold, it indicates that the sampling distribution matches the perceptual information well, and the importance resampling is performed; otherwise, it indicates that the matching with the perceptual information is poor, and the resampling from the perceptual information is performed. For the selection of the threshold, factors such as the perception model and the number of currently observed features should be considered.

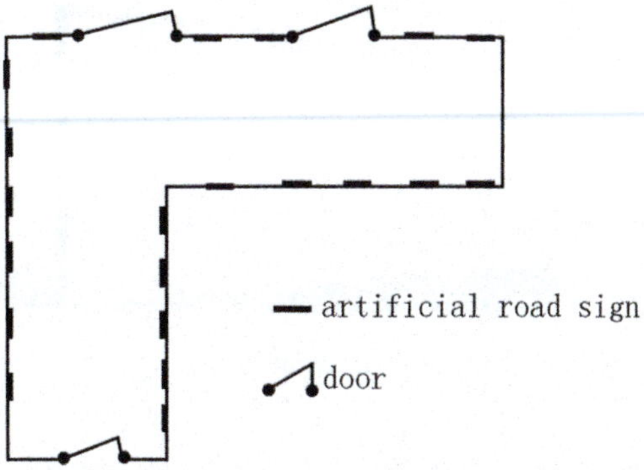

Fig. 4.28 Plane projection of the test environment

Fig. 4.29 Service robot observes a road sign up close

4.5.2 Robot Experiment Environment

The experimental environment is selected as a corridor outside the office, and the plane projection diagram is shown in Fig. 4.28. Because it mainly debugs the positioning function of the mobile robot, based on the principle of convenient debugging, unnecessary equipment such as the casing of the mobile robot is not equipped. Figure 4.29 shows the service robot observing the road sign at close range.

The image captured by the equipped stereo vision system is shown in Fig. 4.30. The stereo vision system we are equipped with has a viewing angle of 45°, and the recognizable features in the environment are pre-set road signs.

4.5.3 Positioning Error Experiment

In the experiment, the mobile robot roams autonomously in the environment. The precision test experiment of robot self-positioning is mainly based on visual sensor, supplemented by odometer sensor. The experimental environment is shown in 4.5.1. Place the road signs at an equal distance in the environment. The red line in Fig. 4.31

Fig. 4.30 The camera takes the image

is the top view of the road signs. The robot roams in a predetermined environment and locates the robot body through the detected road signs. The whole experimental process is given below.

During the experiment, the sampling number is variable. The experimental parameters used by the extended Monte Carlo method are: the effective sampling number threshold $k = 10\%$, the constant c is 0.8, λ is 0.15–0.25, and decreases with the increase of entropy. The proportional coefficient k_w is taken as 50%. For the Monte Carlo method of uniform resampling, since the random sampling introduced will increase the uncertainty of the sampling distribution, the corresponding parameters are different, k remains unchanged, c takes 0.3, λ takes 0.35, and k_w takes 30%.

In order to verify the localization accuracy of the extended Monte Carlo method, the same perceptual model and the same number of samples are used by uniformly resampling the Monte Carlo method and the extended Monte Carlo method. Figure 4.32 shows the error comparison of the robot using two positioning methods. The real pose is obtained through the one-way meter information. Since the ground

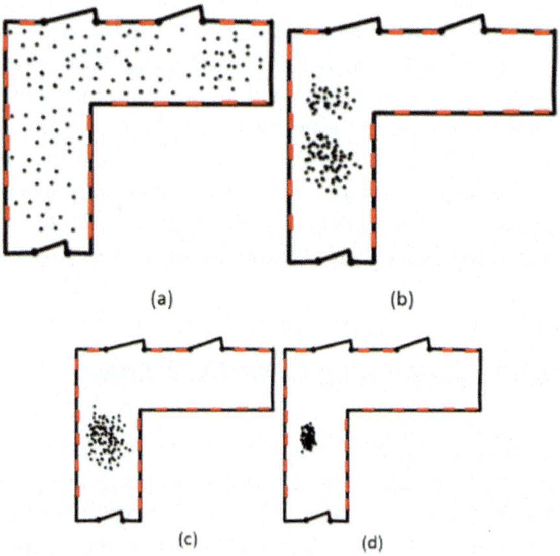

Fig. 4.31 Particle filter positioning process picture, **a**, **b**, **c**, and **d** are the results of particle filtering iteration 1, 5, 12, and 20 times respectively

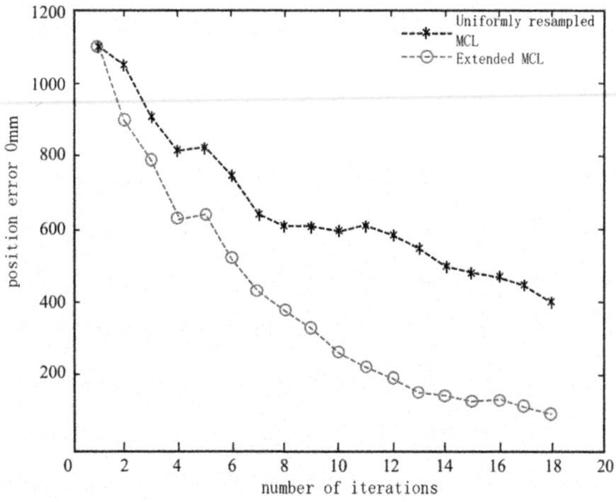

Fig. 4.32 Comparison of positioning algorithm accuracy

in the environment is relatively smooth and the moving distance is shorter, the one-way meter information is more accurate. It can be seen that the resampling-based extended Monte Carlo method is significantly better than the uniform resampling Monte Carlo method in terms of positioning error and convergence speed after being kidnapped.

According to the results of the positioning error, it can be seen that although the positioning precision has been greatly improved, the positioning error fluctuates around 9 cm, but the error is relatively large. For example, the industrial robot vision system, the positioning precision can be accurate to several millimeters about.

There are three main reasons for the error: The first is the application environment. Industrial robots are generally used in relatively fixed occasions. The camera generally does not need to move. Each time an image is acquired, it is generally at a fixed angle and a fixed distance, and the positioning precision is very high. The vision system of the intelligent mobile robot is installed on the robot body, and the angle and distance of each observation environment are random. When the angle of the observation target is good, a relatively accurate positioning result can be obtained. When the observation angle or distance in an unsatisfactory situation, the positioning error will increase relatively; the second reason is the precision of the vision system. The size of the image collected by the stereo vision system we use is 752×480. When the target is observed at a distance of more than 4 m, the actual length represented by each pixel exceeds 1 cm. Pixel errors in edge detection and fitting during image processing are unavoidable, so the precision of the vision system directly affects the final positioning precision; the third reason is the effective road sign detection link of the positioning algorithm. At present, in order to increase the proportion of the vision system in the positioning system, all road signs that can be detected in the field of vision are regarded as valid road signs, and all participate in the calculation of the final positioning of the robot, and some road signs with unsatisfactory angles and distances will bring some errors.

Chapter 5
Path Planning for Mobile Robots Based on Algorithm Fusion

The path planning problem of mobile robots is one of the hot issues in the field of mobile robot research. To find an optimal or near-optimal path from the starting state to the target state in the motion space that can avoid obstacles based on some or some optimization criteria (e.g., shortest travel route, shortest travel time, etc.) is what we call the path planning problem for mobile robots.

According to the scope of application of path planning methods can be divided into global path planning methods, local path planning methods and hybrid path planning methods.

The global planning method is a path planning method applicable with a priori maps, which plans a collision-free optimal path for the robot based on known map information. The degree of perception of the environmental information will determine the accuracy of the planned path since it is highly dependent on the environmental information. Global methods can usually find the optimal solution, but they require accurate information about the environment to be known in advance and are computationally large.

Local path planning is mainly based on the information sensed by the robot's sensors at the current moment for autonomous obstacle avoidance. Among the existing research results, most of the navigation results are local path planning methods, which only need to obtain the current environment information through the sensors they carry, and this information can be updated with the changes of the environment. Compared with global planning methods, local path planning methods are more advantageous in terms of real-time and practicality. The local path planning method also has the drawback that it has no global information and is prone to local poles, which cannot guarantee that the robot can reach the destination successfully.

Since neither global planning nor local planning alone can achieve satisfactory results, a hybrid algorithm that combines the advantages of both has arisen. This method uses the global information of global planning as the a priori condition of

The original version of this chapter was revised: Equation 5.10 has been corrected. The correction to this chapter is available at https://doi.org/10.1007/978-981-99-0589-8_8

T. Guo et al., *Special Robot Technology*, Advanced and Intelligent Manufacturing in China, https://doi.org/10.1007/978-981-99-0589-8_5

local planning to avoid local minima due to the lack of global information in local planning, and thus guides the robot to find the target point eventually.

A good path planning method should not only meet the requirements of rational and real-time path planning, but also satisfy the requirement of being optimal under a certain rule, as well as having the ability to adapt to dynamic changes in the environment.

5.1 Common Path Planning Methods

Currently, the commonly used path planning methods for mobile robots include artificial potential field method, A* algorithm, neural network, fuzzy inference, genetic algorithm and ant colony algorithm.

- Artificial potential field method

The artificial potential field method is a virtual force method proposed by Khatib. The basic idea is to design the motion of the robot in the surrounding environment as an abstract artificial gravitational field, where the target point exerts "gravitational force" on the mobile robot and the obstacle exerts "repulsive force" on the mobile robot, and finally control the motion of the mobile robot by finding the combined force. The motion of the mobile robot is controlled by finding the combined force. The artificial potential field method is simple in structure, easy to control in real time at the bottom, and the planned path is generally smooth and safe, but this method has the problems of local minima and unreachable targets.

- A* algorithm

The path planning problem for mobile robots is a problem solver. Search algorithms are usually used to solve this type of problems. One of the most commonly used path search algorithms is the A* algorithm, which is the most efficient search method for solving the shortest path in a static road network and an effective algorithm for solving many search problems. It is a widely used heuristic search algorithm, which is obtained by continuously searching for paths that approximate the destination. It is based on symbols and logic, and takes as its study how to find a sequence of actions to reach its target bit when no separate action of the intelligence can solve the problem. It is very fast in finding the shortest path (exactly the path with the lowest time cost) on a completely known and relatively simple map, and the search size can be easily controlled using the A* algorithm to prevent blocking. The classical A* algorithm is an extremely effective method for solving shortest paths in static environments.

- Neural network method

The artificial neural network method is a nonlinear dynamical system that simulates human thinking based on the knowledge and understanding of the organization and operation mechanism of the human brain, featuring distributed storage of information, parallel collaborative processing and good self-organizing and self-learning capabilities. It can use environmental obstacles, etc. as the input layer information of the neural network, which is processed in parallel by the neural network, and the output layer of the neural network outputs the desired steering angle and speed, etc., to guide the robot to avoid obstacles and drive until it reaches the destination. The disadvantage of this method is that it must relearn when the environment changes, which is difficult to apply when the environmental information is incomplete or the environment changes frequently.

- Fuzzy reasoning method

Fuzzy theory was developed on the mathematical basis of fuzzy set theory founded by Professor L. A. Zadeh of the Department of Electrical Engineering at the University of California, Berkeley, and mainly includes aspects of fuzzy set theory, fuzzy logic, fuzzy inference and fuzzy control.

The human driving process is essentially a fuzzy control behavior, the size of the curvature of the path, the size of the position and directional deviation are fuzzy quantities obtained by the human eye, and the driver's driving experience is impossible to determine precisely, fuzzy control is an effective way to solve this problem. Mobile robots are similar to vehicles in that their kinematic models are more complex and difficult to determine, while fuzzy control does not require an accurate mathematical model of the control system. In addition, a mobile robot is a typical time-delayed, nonlinear unstable system, and a fuzzy controller can accomplish a nonlinear mapping from the input space to the output space.

Fuzzy theory for mobile robot path planning combines the robustness of fuzzy reasoning with the physiologically based "perception–action" behavior to quickly reason about obstacles with good real-time performance. The method avoids the disadvantages of other algorithms, such as the strong dependence on environmental information, and shows great superiority and high real-time performance for robot path planning in complex environments.

- Genetic algorithm

Genetic algorithm is a randomized search algorithm that draws on natural selection and natural genetic mechanisms in biology. Due to its advantages of robustness and global optimization, it has good applicability to complex and nonlinear problems that are difficult to be solved by traditional search methods. Applying genetic algorithm to solve the problem of obstacle avoidance and path planning in dynamic environment of mobile robot can avoid difficult theoretical derivation and obtain the optimal solution of the problem directly. However, there are also some shortcomings, such as not fast computation speed and early convergence.

- Ant colony algorithm

Ant colony algorithm, is a kind of optimization algorithm used to solve complex problems. Ant colony algorithm was early proposed to solve the travel salesman problem. With the in-depth study of ant colony algorithm, it was found that ant colony algorithm has a wide application prospect in solving secondary optimization problems, so ant colony algorithm also gradually developed to more fields from solving TSP problems in the early days. At present, the ant colony algorithm has a wide range of applications in solving scheduling problems, bus route planning problems, robot path selection problems, network routing problems, and even in enterprise management problems, pattern recognition and image alignment.

Ant colony algorithms are not only capable of intelligent search and global optimization, but also have features such as robustness, positive feedback, distributed computation, easy integration with other algorithms, and constructiveness, and can add features to artificial ant colonies that are not found in natural ant colonies, such as foresight and backtracking, as needed.

Although the ant colony algorithm has many advantages, the algorithm also has some drawbacks. Compared with other methods, the algorithm generally requires a longer search time, and although the increase in computer computing speed and the essential parallelism of the ant colony algorithm can alleviate this problem to some extent, it is a major obstacle for large-scale optimization problems. Moreover, the method is prone to stagnation, i.e., after the search proceeds to a certain extent, when the solutions found by all individuals converge, no further search of the solution space can be performed, which is not conducive to the discovery of better solutions. In the ant colony system, ants always rely on the feedback information from other ants to reinforce learning without considering their own experience accumulation, such a blind obedience behavior, which easily leads to premature and stagnation phenomenon, thus slowing down the convergence speed of the algorithm. Based on this, scholars have proposed improvement algorithms for ant colony systems. For example, Dorigo M. et al. proposed an ant colony algorithm called Ant-Q System; German scholars Stutzle T. and Hoos H. proposed the MAX–MIN Ant System (MMAS). Wu Q. H. et al. were inspired by the role of variation operator in the genetic algorithm, and adopted the mechanism of reversing variation in the ant colony algorithm, and then proposed an ant colony algorithm with variation characteristics. Since then, improved ant colony algorithms have been proposed, such as ant colony algorithm with sensory and perceptual features, adaptive ant colony algorithm, ant colony algorithm based on pheromone diffusion, ant colony algorithm based on hybrid behavior, and ant colony algorithm with small windows based on pattern learning.

5.2 Path Planning Based on Fusion of Artificial Potential Field and A* Algorithm

5.2.1 Artificial Potential Field Method

The gravitational and repulsive force distributions of the artificial potential field method are shown in Fig. 5.1. Where, F_{att} is the gravitational force of the robot by the target point and F_{rep} is the repulsive force of the robot by the obstacle. F_{total} is the combined force generated by the gravitational and repulsive forces, which control the robot's motion toward the target point.

The mathematical description of the artificial potential field method is as follows.

Let the current position information of the robot be $R = (x, y)$, and the position information of the target point be $R_{goal} = (x_{goal}, y_{goal})$, the target G acts as an attraction to the mobile robot R, and the further the distance, the greater the attraction, and vice versa, the smaller the attraction.

The gravitational field between the robot and the target point is defined as

$$U_{att} = \frac{1}{2} k_{att} \rho(R, R_{goal})^2 \tag{5.1}$$

where k_{att} is the gravitational field gain factor; $\rho(R, R_{goal})$ is the distance between the robot's current position R and the target point R_{goal}.

The gravitational force on the robot generated by this gravitational field is a negative gradient function of the gravitational potential energy on the robot.

$$F_{att}(R) = -\nabla U_{att}(R) = -k_{att} \rho(R, R_{goal}) \tag{5.2}$$

where the gravitational force $F_{att}(R)$ is directed on the line between the robot and the target point, from the robot to the target point. This gravitational force tends to

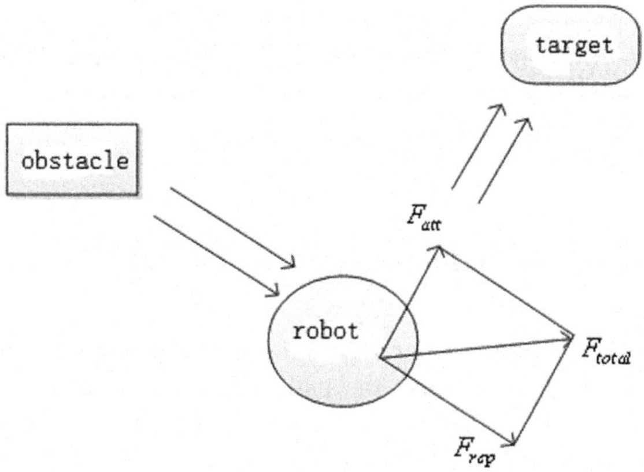

Fig. 5.1 Force analysis diagram of the mobile robot based on the artificial potential field method

zero linearly as the robot tends to the target, and is zero when the robot reaches the target point.

The repulsive field function is formulated as follows

$$U_{att}(R) = \begin{cases} \frac{1}{2}k_{rep}\left(\frac{1}{\rho(R,R_{obs})} - \frac{1}{\rho_0}\right)^2 & , \rho(R, R_{obs}) \le \rho_0 \\ 0 & , \rho(R, R_{obs}) > \rho_0 \end{cases} \tag{5.3}$$

where, k_{rep} is the positive scale factor, $R_{obs} = (x_{obs}, y_{obs})$ is the position of the obstacle, $\rho(R, R_{obs})$ is the distance between the robot and the obstacle, and ρ_0 indicates the influence distance of the obstacle. The repulsive force generated by this repulsive site is the negative gradient of the repulsive potential energy.

$$\begin{aligned} F_{rep}(R) &= -\nabla U_{rep}(R) \\ &= \begin{cases} k_{rep}\left(\frac{1}{\rho(R,R_{obs})} - \frac{1}{\rho_0}\right)\frac{1}{\rho^2(R,R_{obs})}\nabla\rho(R, R_{obs}), & \rho(R, R_{obs}) \le \rho_0 \\ 0 & , \rho(R, R_{obs}) > \rho_0 \end{cases} \end{aligned} \tag{5.4}$$

The combined force on the robot is equal to the sum of the gravitational force and the repulsive force, i.e.

$$F_{total} = F_{att} + F_{rep} \tag{5.5}$$

The artificial potential field method is simple in structure and has good performance in real-time control of the robot trajectory, but according to the above principle, it is only suitable for solving the local obstacle avoidance problem, and in some areas of the global map, when the robot is subject to the joint action of the gravitational potential field function and the repulsive potential field function, the robot is easy to produce oscillation or stagnation at a certain location, which is the so-called local minimal point. The probability of generating local minima is proportional to the number of obstacles, and the more obstacles there are, the higher the probability of generating local minima.

5.2.2 A* Algorithm

A* algorithm, a new type of heuristic search algorithm, has been applied by many researchers to solve the path planning problem of mobile robots because of its advantages such as rapid search and easy implementation.

The basic idea of A* search is that if C_S is the starting point and C_G is the target point, then for any point in the environment C_i, assume that $g(C_i)$ denotes the minimum path cost from C_S to C_i and $h(C_i)$ denotes the estimated cost from C_i to C_G, giving the estimated cost of the optimal path from the initial point C_S through the intermediate point C_i to the target point C_G, represented by a heuristic evaluation function $f(C_i)$, i.e.

$$f(C_i) = g(C_i) + h(C_i) \tag{5.6}$$

In a dynamic environment, the A* algorithm first plans an optimal path from the initial location to the target location based on known environmental information, and then the robot walks along this path through motion control. When the robot senses that the current environmental information does not match the known environmental map, it models the current environmental information and updates the map, and the robot replans the path according to the updated map. However, if the robot is in an environment where there is no a priori map or the environmental information is constantly changing, the robot has to replan the path very frequently, and this replanning algorithm is also a global search process, which will increase the amount of system operations. If the A* method uses a raster representation of the map, the accuracy of the map environment representation increases as the raster size decreases, but at the same time the search range of the algorithm increases exponentially, if the raster size increases, the search range of the algorithm decreases, but the accuracy and success rate of the algorithm decreases. The A* method is optimized by improving the local path planning method of artificial potential field, which can effectively increase the raster size of A* method and reduce the computational power of A* method.

5.2.3 *Artificial Potential Field and A* Algorithm Fusion*

This study combines the A* algorithm and the artificial potential field algorithm to propose a path planning method that can integrate the global path planning method and the local path planning method.

- Description of the fusion of path planning algorithms

If the information of the starting point of the mobile robot is denoted by S, the information of the target point of the robot is denoted by G, the information of the current position of the robot is denoted by C, and the information of the raster environment map is denoted by M, then the hybrid path planning method proposed in this study can be specifically described as

Step 1: Rasterize the current perceived and known environmental information of the robot and save it in the raster map M. Assign the robot's starting point state to the current position state, i.e., C = S.

Step 2: The mobile robot plans a global optimal path from the current location C to the target point G based on the saved raster map M, generating a sequence of sub-target nodes; if there is no information in this sequence, which means there is no feasible path at present, the search is returned as a failure.
Step 3: Determine the nearest sub-target node to the current location.
Step 4: Update the target point of the system, use the sub-master node obtained in step 3 as the new target point G_1, and perform motion control according to the local path planning method until it reaches the position where the target point G_1 is located, and go to step 3.
Step 5: If the sensor carried by the robot itself senses a mismatch between the new environment and the original map M, it updates the map M according to the sensor information, so that C = S, and jumps to the first step.

In the above algorithm, the generation of the sequence of sub-target nodes is implemented in the global path planning module, and the control of the mobile robot is implemented in the local path planning module, and it keeps moving toward the sub-target nodes while updating the sub-target nodes, and finally reaches the final target point.

The following describes global path planning and local path planning respectively.

- Global path planning method

Global path planning uses the raster map-based A* search method for path planning. In the global path planning, the dynamic obstacle avoidance of the robot is not considered, and the granularity of the raster can be set larger, which can reduce the use of system space as well as reduce the computation of A* search and improve the efficiency of A* search.

When the A* algorithm is used for global path planning, the global map as well as the local map is rasterized firstly, a raster coordinate system is established, and the raster coordinates of the initial and target points are obtained by transforming the global and raster coordinates. In the raster coordinate system, the A* algorithm searches for an optimal path from the starting point to the target point and generates a two-dimensional sequence of sub-target points. The information stored in each sub-target point in the sequence is the raster coordinates of its location. In this set of sequences, each node except the global target node has a pointer to its parent node. Then the coordinates of the point under the global coordinates are obtained by transforming the raster coordinates of each child target node and the global coordinates. If the robot is at any position other than the grid where the target endpoint is located, the robot is still subject to the gravitational force generated by the parent node of the grid where it is located, and when the robot reaches the grid where the target endpoint is located, the robot is subject to the gravitational force of the robot's target node.

The search is performed by the A* algorithm, and all that is obtained is a sequence of sub-target nodes. In the following, a local path planning algorithm will be used to realize the smooth motion of the robot following the above path.

- Local path planning method

In this study, an improved artificial potential field method is used for local path planning of mobile robots. The artificial potential field method is improved in terms of two aspects: obstacle area and kinematic control constraints.

– Setting up an effective obstacle that generates a repulsive force function

During the actual motion of the mobile robot, only a limited number of obstacles can exert repulsive force on the robot, and only obstacles within a certain range from the robot's direction of motion will have an impact on the robot's motion. In the local path planning, it is assumed that the angle between the robot's forward direction and the obstacles is α, and the obstacles are distributed as shown in Fig. 5.2. In the improved artificial potential field method, only obstacles 1 and 2 within a certain range in the positive direction of robot motion will have a repulsive potential field on the robot, and obstacles in other directions will not affect the robot motion.

The repulsive force function on the robot under the repulsive field of obstacle 1 is as follows.

$$F_{rep1}(R) = -\nabla U_{rep1}(R) = \begin{cases} k_{rep1}\left(\frac{1}{\rho(R, R_{obs1})} - \frac{1}{\rho_1}\right)\frac{1}{\rho^2(R, R_{obs1})}\nabla\rho(R, R_{obs1}), & \rho(R, R_{obs1}) \leq \rho_1 \\ 0, & \rho(R, R_{obs1}) > \rho_1 \end{cases} \tag{5.7}$$

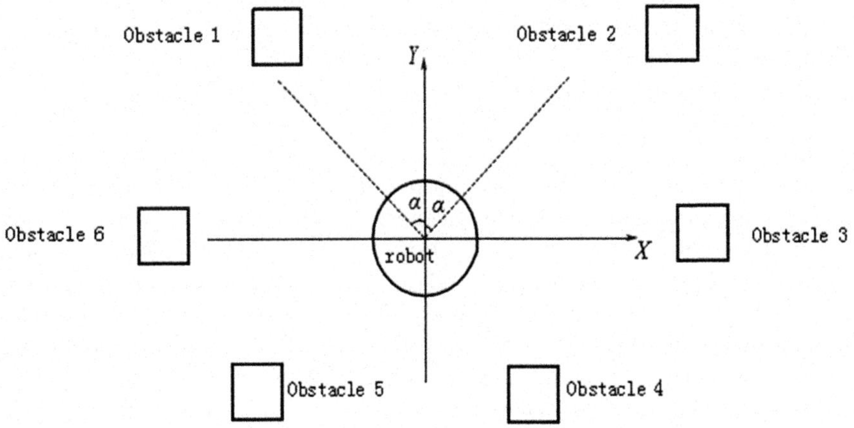

Fig. 5.2 Obstacle distribution around the robot

where, k_{rep1} is the positive scale factor, $R_{obs1} = (x_{obs1}, y_{obs1})$ is the position of obstacle 1, $\rho(R, R_{obs1})$ is the distance between the robot and obstacle 1, and ρ_1 indicates the influence distance of the obstacle.

The repulsive force function on the robot under the repulsive field of obstacle 2 is as follows.

$$F_{rep2}(R) = -\nabla U_{rep2}(R)$$
$$= \begin{cases} k_{rep2}\left(\frac{1}{\rho(R,R_{obs2})} - \frac{1}{\rho_2}\right)\frac{1}{\rho^2(R,R_{obs2})}\nabla\rho(R, R_{obs2}), & \rho(R, R_{obs2}) \le \rho_2 \\ 0 & , \rho(R, R_{obs2}) > \rho_2 \end{cases} \tag{5.8}$$

where, k_{rep2} is the positive scale factor, $R_{obs2} = (x_{obs2}, y_{obs2})$ is the position of the obstacle, $\rho(R, R_{obs2})$ is the distance between the robot and the obstacle, and ρ_2 indicates the influence distance of the obstacle.

The combined repulsive force on the robot in the region of the cluster of obstacles is

$$F_{rep} = F_{rep1} + F_{rep2} \tag{5.9}$$

Using this method, not only can the efficiency of local path planning be improved, but also can effectively reduce the local minimal point problem generated by the artificial potential field method, so that the robot can cross the multi-obstacle area quickly and safely. When the robot reaches the grid where the target point is located, then the robot will no longer be influenced by the surrounding environment, and it is subject to the attraction generated by the target endpoint to it, and the problem of unreachability of the target point near the obstacles can be solved.

Obstacle distribution around the robot, as shown in Fig. 5.2.

- Kinematic control constraints

After generating the angular and linear velocities of the robot motion by the above local path planning, the motion control module converts the obtained linear and angular velocities into the velocities of the left and right wheels of the mobile robot that the motors can recognize v_l and v_r, and then performs trapezoidal planning for the velocities of the left and right wheels, so that the wheel speeds can be smoothly incremented or decremented to prevent the sudden changes in wheel speeds from causing overshoot of the motion control. It is also necessary to limit the calculated speed in order to ensure the safety of the motor.

5.2.4 Simulation Research

Figure 5.3 shows the simulated path planning of the robot from one room to another. The start point is START and the goal point is GOAL.

As seen in Fig. 5.3a–d, the robot completes the process from one room (starting point) to another room (target point), and the global path planner generates the sub-target sequence points of the global optimal path in the current environment, and the improved artificial potential field method is used to control the robot's motion between the sub-target sequence points. Eventually, the robot is able to move from the initial point to the target point along a smooth path.

Figure 5.4 shows the simulation diagram of the robot at the same initial position and at two different target positions, respectively. From the comparison of Fig. 5.4a, b), it can be found that the robot is always able to plan a trajectory that is optimal (shortest path) from the initial point to the target point in a global sense for the same initial position of the robot with different target positions.

Figure 5.5 shows the simulation effect of path planning based on artificial potential field and A* fusion algorithm for a robot in a complex environment.

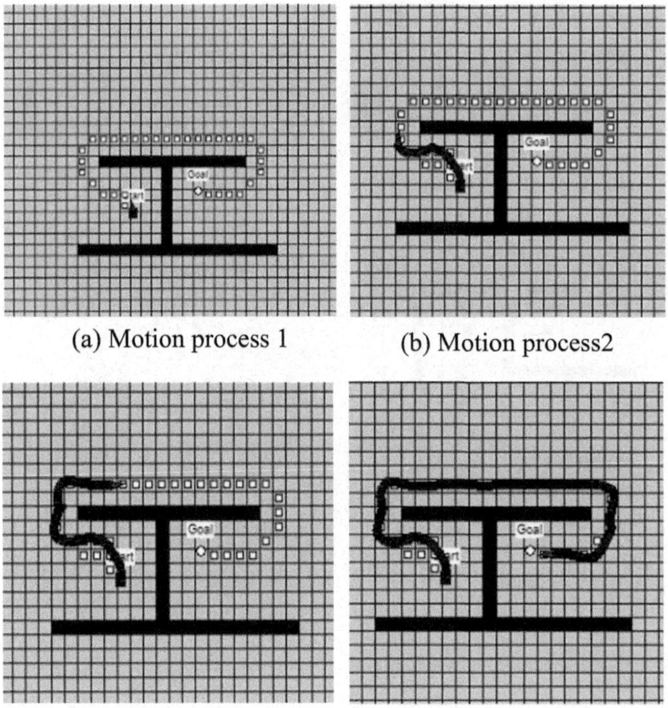

(a) Motion process 1 (b) Motion process2

(c) Motion process3 (d) Motion process4

Fig. 5.3 Simulation effect of the robot's movement from the starting point to the target point

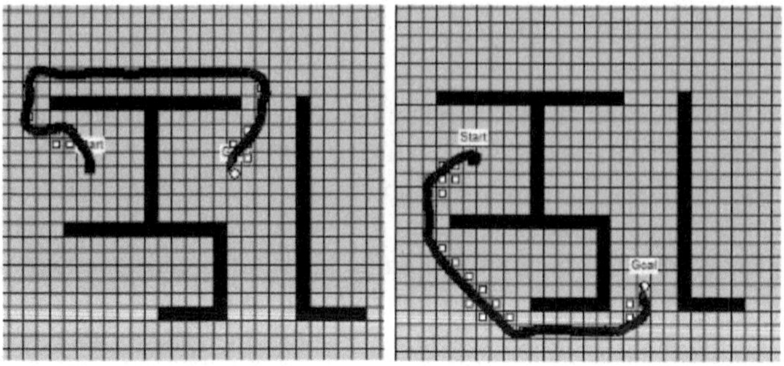

(a) Target position 1 (b) Target position 2

Fig. 5.4 Global path planning simulation comparison chart

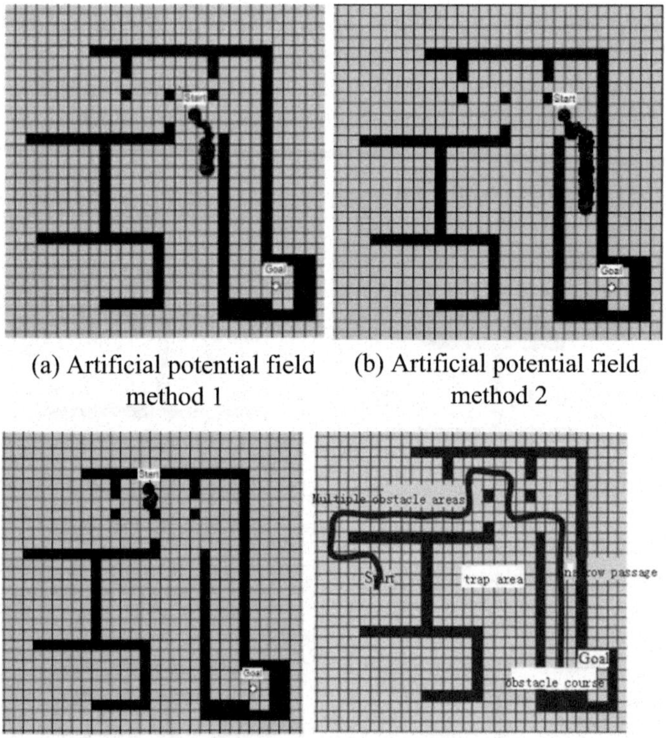

(a) Artificial potential field method 1 (b) Artificial potential field method 2

(c) Artificial potential field method 3 (d) Artificial potential field method 4

Fig. 5.5 Global path planning simulation effect

As can be seen in Fig. 5.5(a–c), in the artificial potential field method path planning, there are trap areas, unrecognizable robot paths in multiple obstacle areas and oscillation problems in narrow passages. As can be seen in Fig. 5.5(d), using the fusion algorithm proposed in this study, the robot is able to effectively avoid trap regions throughout its motion according to the guidance of the sequence of sub-target nodes planned out in the global path; when the robot traverses a multi-obstacle region, only the obstacles that can affect the robot's motion need to be considered under the guidance of the sub-target nodes, so as to avoid the occurrence of oscillation. When the robot moves through the narrow passage, it can pass the passage smoothly; when the robot reaches the target node near the obstacle, it ignores the influence caused by the environmental information at this time, and the robot can reach the target point successfully. It can be seen from this simulation that the improvement of the artificial potential field method in this study can well solve the defects of the classical artificial potential field method and overcome the problem of unreachability of the target point near the obstacle. Meanwhile, it can be seen from the simulation experiments that the robot always has a certain distance from the obstacle during the motion, which can ensure the safety of the robot in the motion and avoid the local speed overshoot causing the robot to collide with the obstacle. Figure 5.6 simulate the process of the robot encountering dynamic obstacles in the path planning and avoiding them effectively.

As can be seen in Fig. 5.6(a–b), the robot follows the currently planned path toward the target point, and when the robot enters the grid region where the sub-target point shown in Fig. 5.6(c) is located, the robot detects dynamic obstacles at this time, and the robot is able to effectively avoid dynamic obstacles using the algorithm in this paper, and the robot continues to tend to move toward the target node as shown in Fig. 5.6(d). From the above simulation results, it can be seen that due to the local path planning strategy, the robot can avoid obstacles in dynamic environment in real time, meet the global optimum of path planning performance, and have excellent adaptability to dynamic environment, which is very suitable for the application of mobile robot path planning in complex environment.

5.3 Robot Path Planning Based on Artificial Potential Field and Ant Colony Algorithm

Ant colony algorithm is an optimization algorithm that simulates the behavior of ant colony proposed by Italian scholars Dorigo et al. in the late twentieth century. The algorithm is a combination of distributed computation, positive feedback mechanism and greedy search, with a strong ability to search for better solutions. Positive feedback can find the better solution quickly, distributed computing avoids premature convergence, and greedy search helps to identify acceptable solutions early in the search process, shortens the search time, and has strong parallelism. However, ant colony algorithms generally require more search time and are prone to stagnation,

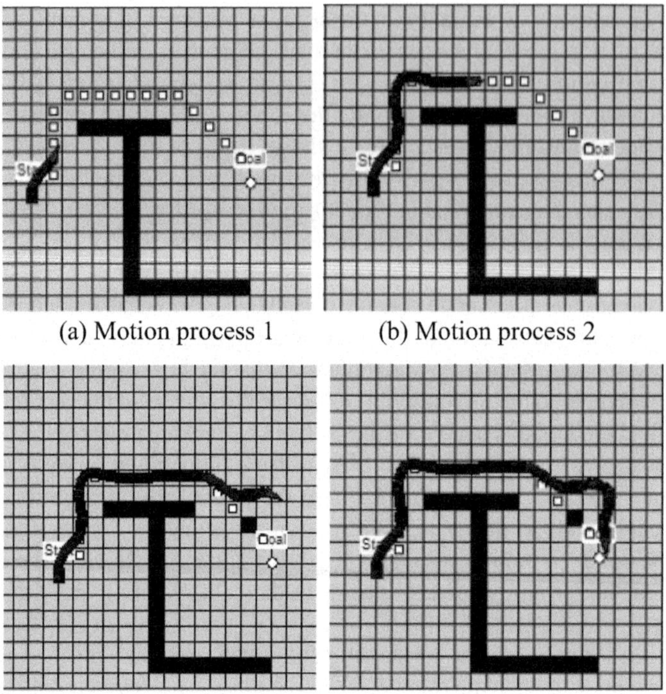

(a) Motion process 1 (b) Motion process 2

(c) Motion process 3 (d) Motion process 4

Fig. 5.6 Simulation effect of local path planning and obstacle avoidance

which is not conducive to finding better solutions. Artificial potential field method is a more mature local path planning method with simple structure and small computation, which can quickly drive the robot toward the target point by potential field force, but it is easy to fall into local minimal points and oscillate in front of obstacles.

In order to better solve the path planning problem of mobile robots, this study proposes a potential field ant colony algorithm that combines the potential field force of the artificial potential field method with the ant colony algorithm, and avoids the "blind search" of individual ants at the beginning of the ant colony algorithm by adding the potential field force, and suppresses the influence of the potential field force on the later stage of the ant colony algorithm by adding the restriction condition.

5.3.1 Ant Colony Algorithm

The ant colony algorithm is a bionic algorithm that uses the strength of the pheromone left behind by the route taken by simulated ants to the target point to achieve optimal path planning.

Let m denote the number of ants; $d_{ij}(i, j = 0, 1, 2 \cdots\cdots n)$ denotes the distance between node i and node j, where n is the number of nodes; α is the pheromone factor, which denotes the relative importance of trajectories; β is the expectation heuristic factor, which denotes the relative importance of visibility; ρ is the pheromone volatility factor, and $0 \leq \rho < 1$; the initial number of iterations N = 0 and the maximum number of iterations is $N_{\max}$.

The specific implementation steps of the ant colony algorithm path planning are as follows.

Step 1: At the initial moment, m ants are randomly placed on the grid map and the initial pheromone concentration is the same on each path. The grid of obstacles is denoted by 0. The grid that allows the robot to perform movement is set to 1, and the parameters are initialized.

Step 2: If we let $\tau_{ij}(t)$ be the residual pheromone concentration between two nodes i, j at moment t; $\eta_{ij}(t)$ be the expectation heuristic function between two nodes i, j at moment t, defined as the reciprocal of the distance d_{ij} between nodes i and j;, $Tabu_k(k = 1, 2, \cdots, m)$ is the set of nodes that ant k has traveled, $allowed_k = \{1, 2, \cdots, n - Tabu_k\}$ indicates the set of nodes are not in $Tabu_k$, that is, the set of nodes that the ant can choose next. Then, the state transfer probability of ant k moving from node i to node j at moment t is

$$P_{ij}^{k}(t) = \begin{cases} \dfrac{\tau_{ik}^{\alpha}\eta_{ik}^{\beta}}{\sum\limits_{s \in allowed_k} \tau_{ik}^{\alpha}\eta_{ik}^{\beta}} & s \in allowed_k \\ 0 & \text{otherwise} \end{cases} \tag{5.10}$$

By using the above state transfer formula, the grid to which the ant will be transferred next is calculated, and the grid of the obstacle is not walkable.

Step 3: Pheromones are left on the paths taken by ants, and to avoid overwhelming the inspired information due to too many residual pheromones on the paths, the pheromones are volatilized with the passage of time. The pheromone update rules on nodes i and j at $t+\Delta t$ are shown in Eq. (5.11), and then the length of the shortest path taken by ants in this iteration is saved and the next iteration is started.

$$\begin{cases} \tau_{ij}(t + \Delta t) = (1 - \rho) \cdot \tau_{ij}(t) + \Delta\tau_{ij}(t) \\ \Delta\tau_{ij}(t) = \sum\limits_{k=1}^{m} \Delta\tau_{ij}^{k}(t) \end{cases} \tag{5.11}$$

If we let Q denote the total amount of pheromones secreted by ant in this cycle, L_k is the total length of the path taken by ant k in this cycle, and $p_k(begin, end)$ is the path taken by ant k from the starting point to the end point in this cycle, using the Ant-Cycle model, then we have

$$\Delta\tau_{ij}^{k}(t) = \begin{cases} \frac{Q}{L_k} & (i, j) \in p_k(begin, end) \\ 0 & \text{otherwise} \end{cases} \tag{5.12}$$

and save the length of the shortest path taken by the ant in this iteration and start the next iteration.

In the new iteration, ants can walk by the pheromone left by the shortest path of the last ant. The grids containing pheromones are more likely to be selected by the current ant state shift. Therefore, after each iteration, the ant will search for a closer path around the nearest path.

Step 4: Repeat step 2 and 3 to determine if the maximum number of iterations is reached, and stop iteration if it is reached. Take the path which has the shortest distance as the output item.

5.3.2 *Improved Artificial Potential Field Method*

The traditional artificial potential field method has the problems of easily falling into the local minima and wandering in front of the concave obstacles. The artificial potential field method is improved by adding the filling field to solve the problem of the robot wandering in front of obstacles and falling into local minimum points to some extent. The specific steps of the method are as follows.

- Determine whether the robot's current pose and the previous n-step pose are repeatedly changing within a small range, and if they are essentially constant within a small range, identify obstacle avoidance difficulties.
- The robot makes a retreat and exits outside the area of difficulty in obstacle avoidance.
- In the difficult region of obstacle avoidance, a filling potential field U_{att1} is added to the gravitational field of Eq. (5.1) U_{att}, whose expression is

$$U_{att1} = \begin{cases} K\dfrac{1}{\rho^2(R, R_{local})}, & \rho(R, R_{local}) \le \rho_r \\ 0, & \rho(R, R_{local}) \le \rho_r \end{cases} \tag{5.13}$$

$R_{local} = (x_{local}, y_{local})$ is the position of the local minima, $\rho(R, R_{local})$ is the distance between the robot's current position and the local minima, and $K \in R^+$ is the scale factor in the potential field U_{att1}, which is a positive constant. $\rho_r \in R^+$ is the radius within which the potential field U_{att1} can affect the mobile robot.

5.3.3 *Ant Colony Algorithm Based on Potential Field Force Guidance*

There are various ways to combine artificial potential field and ant colony algorithms. In this study, the potential field force in the improved artificial potential field method

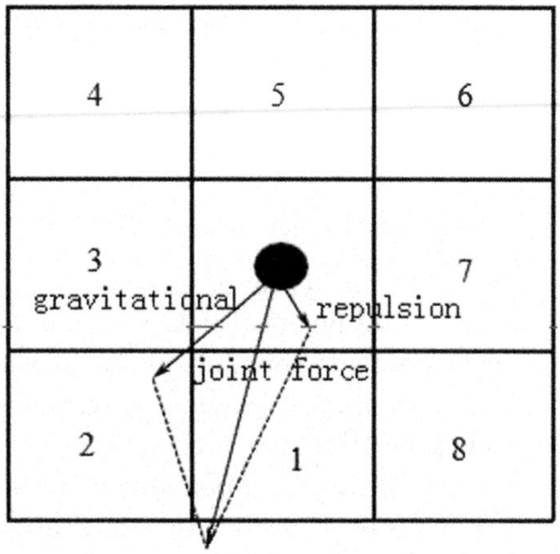

Fig. 5.7 Potential field forces on ants

is combined with the ant colony algorithm, and the combined algorithm makes the ants calculate the combined force generated by the repulsive force F_{rep} and the gravitational force F_{att} once before each grid is ready to go to the next grid. F_{total}, the grid with the combined force gets the larger pheromone weight, and then the weights of the pheromones in the direction of the remaining grids on both sides are decreasing in turn. The range of values is $\rho_{rx} \in (1, 2)$. If X is the ordering of the eight grids around the ant, the potential field force points to direction X as 1, and the pheromone weights are the same when X is 2 and 8, 3 and 7, 4 and 6, as shown in Fig. 5.7. This ensures that the direction pointed by the combined force has the highest probability of being chosen by the ants to walk, and also guarantees that the remaining directions may also be chosen.

Therefore, in the initial stage of the algorithm the influence of potential field forces on the ants is sufficiently large, and in order to limit the influence of potential field forces on the ant colony algorithm at a later stage, the following restriction is added.

$$\mu = \frac{N_{\max} - N_n}{N_{\max}} \tag{5.14}$$

$$\rho_{rx\mu} = \rho_{rx} \cdot \mu \cdot F_{total} \tag{5.15}$$

$$\rho_{rx\mu} = \begin{cases} \rho_{rx\mu} & \rho_{rx\mu} \geq 1 \\ 1 & \rho_{rx\mu} < 1 \end{cases} \tag{5.16}$$

where $N_{\max}$ is the total number of iterations and N_n is the current number of iterations. When $\rho_{rx\mu} < 1$ is used, the potential field force no longer affects the ant colony algorithm. This allows the potential field force to compensate for the shortcomings of the early stages of the ant colony algorithm without affecting the later stages of the ant colony algorithm too much. So the probability formula for ant transfer of the improved algorithm is

$$P_{ij}^{k}(t) = \begin{cases} \dfrac{\tau_{ij}^{\alpha}(t)\eta_{ij}^{\beta}(t)\rho_{rx\mu}}{\sum_{s \in allowed_k} \tau_{ij}^{\alpha}(t)\eta_{ij}^{\beta}(t)}, & s \in allowed_k \\ 0, & otherwise \end{cases} \tag{5.17}$$

The specific implementation steps of the algorithm are as follows.

Step 1: Parameter initialization.
Step 2: Add potential field forces to the ant colony algorithm so that each ant calculates the potential field forces when it chooses the direction of movement, calculates the gravitational and repulsive forces on the ants at each node in that path, and derives the direction in which the combined forces point.
Step 3: The transfer probability is calculated by Eq. (5.17), which enables the ants to transfer between the grids they can walk on.
Step 4: Determine whether the distance between the current position of the walking ant and the position of the previous n steps is within a certain threshold, if so, backtrack and re-calculate and add to the taboo table, if not, continue to move forward.
Step 5: Determine whether the ant currently walking has reached the end point, and if it has not reached the end point, continue to calculate and plan the forward direction according to step 3. Otherwise, update the pheromone left when it walks.
Step 6: Check if all the ants in the colony have finished walking, and if there are still remaining ants that have not completed the path search, go to step 4 to continue the search. When all ants have finished searching, proceed to the next step.
Step 7: Judge whether the termination condition is satisfied, if not complete then go to step 3 to continue solving, if the maximum number of iterations has been reached, then save and output the optimal path.

5.3.4 Simulation Research

In this research, the traditional ant colony algorithm and the ant colony algorithm guided by potential field forces proposed in this study are compared and analyzed.

Figure 5.8 show the graphs of the optimization-seeking paths of the conventional ant colony algorithm and the different methods of the ant colony algorithm guided by potential field forces proposed in this study, where the number of parameter ants is $m = 20$, the heuristic factor $\alpha = 1$, the expectation heuristic factor $\beta = 7$, the volatility factor $\rho = 0.3$, and the maximum number of iterations $N = 100$.

The data in Table 5.1 can be derived from the convergence curves obtained in Fig. 5.9. The results show that the potential field force-guided ant colony algorithm converges faster and has shorter path length than the traditional ant colony algorithm with the same parameters.

The artificial potential field method is combined with the ant colony algorithm, because the pheromone weights only have a greater influence on the ants at the

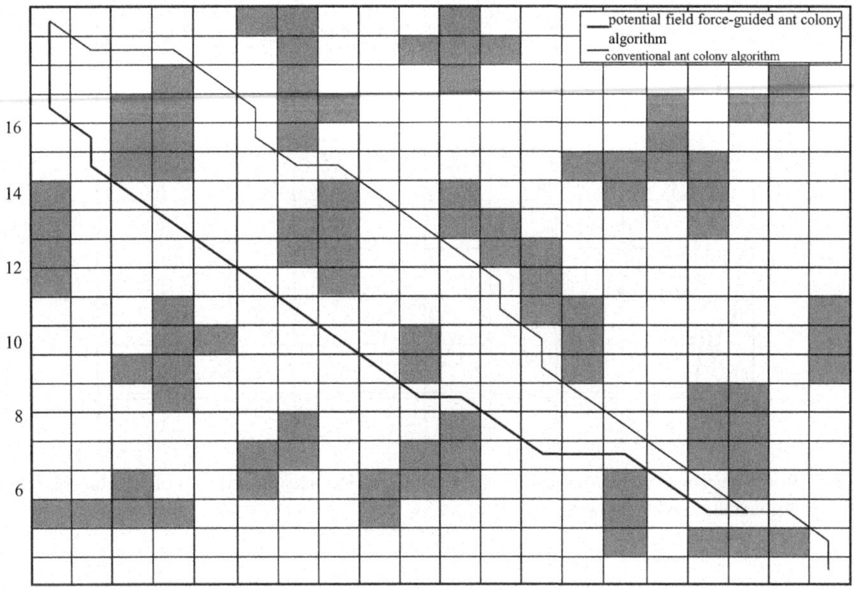

Fig. 5.8 Optimal paths formed by the conventional ant colony algorithm and the potential field force-guided ant colony algorithm, respectively

Table 5.1 Comparison of data of different methods

Algorithm name	Average path length	Optimal path length	Average number of iterations
Traditional ant colony algorithm	31.213	30.627	83
Potential field force guided ant colony algorithm	30.624	29.796	9

beginning, the initial solution of the ant colony algorithm is guided by the potential field force of the artificial potential field method, which reduces the randomness and blindness of the ant colony at the beginning and speeds up the convergence of the ant colony algorithm; and the potential field force on the ants is continuously reduced as the number of iterations increases, the influence of the potential field force on the ants tends to zero in the later stage, in order to better exploit the optimal-seeking ability of the ant colony, and to reduce the possibility of falling into the local optimum to a certain extent through the back-off mechanism. The simulation comparison with the traditional ant colony algorithm shows that the proposed algorithm not only accelerates the convergence speed of ant colony, but also gives full play to the global optimization capability of ant colony.

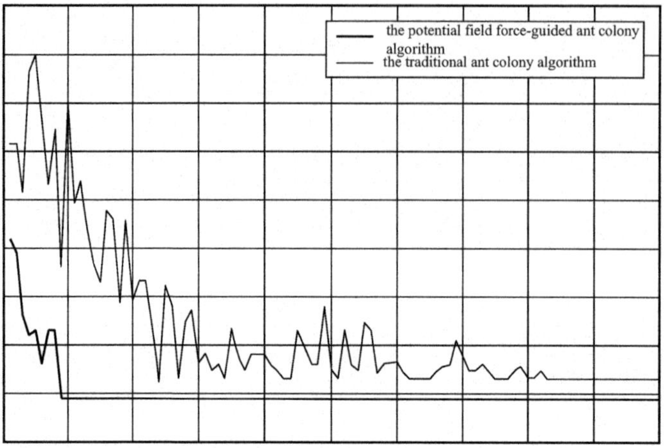

Fig. 5.9 Convergence curves formed by the potential field force-guided ant colony algorithm and the traditional ant colony algorithm

Chapter 6
Ruin Search and Rescue Robot

6.1 Overview of Ruin Search and Rescue Robot

6.1.1 Research Significance of Ruin Search and Rescue Robots

In recent years, earthquake disasters have occurred frequently around the world, not only seriously threatening human safety, but also causing huge economic losses to the affected areas. In 2011, the earthquake and tsunami in northeastern Japan led to a nuclear leak at the Fukushima nuclear power plant with a severity level of 7 (the highest level), and three operators were severely exposed to nuclear radiation during the subsequent nuclear reactor restoration work. Based on the physiological limits of the human body and the harsh survival environment inside the earthquake ruins, 72 hours after the disaster is the golden time for rescue.

More than 72 hours after the disaster, the survivors' lives would be extremely physiologically challenged and their chances of survival would be very slim.

In the earthquake disaster rescue process, the search for survivors inside the rubble is done mainly through the following methods: one is to determine whether there are survivors inside the rubble by listening to the sounds returned from outside the rubble by rescuers asking for them. This method requires survivors to be conscious, and can be used in the early stages of a disaster when the survivors are less severely injured. Another method is to carry professional life detection equipment, rescue workers into the rubble for survivor search work. This method has a large operating range and can also search for survivors with serious injuries or even unconsciousness, but the aftershocks after the earthquake and the poor environment of the rubble can pose a great threat to the safety of the rescuers who enter the rubble, and there have been tragedies in the search and rescue work of major earthquakes in which search and rescue workers have died. With the development of robotics, the application of mobile robots in earthquake rubble search and rescue not only eliminates the risk of rescuers risking their lives to enter the rubble to search for survivors, but also expands

T. Guo et al., *Special Robot Technology*, Advanced and Intelligent Manufacturing in China, https://doi.org/10.1007/978-981-99-0589-8_6

the scope and efficiency of search and rescue activities by carrying professional life detectors into the rubble to search for survivors.

The number of urban population is increasing year by year due to the global wave of urbanization, while natural disasters such as earthquakes lead to more urban population being affected and trapped, and the demand for search and rescue work carried out for the building debris environment is increasing. As a result of disasters, the building debris environment becomes a typical unstructured environment after a disaster. Slipped masonry and gravel cause the ground to become a non-flat environment; collapsed and slipped building structures exacerbate the degree of unstructuredness within building rubble, and at the same time can severely impede the propagation of electromagnetic signals; preserved staircases also severely limit the operational range of search and rescue robots. At present, the motion control of the ruin search and rescue robots mainly adopts two control methods: tele-operated control and autonomous motion control. After entering the ruins, the robot and the controller outside the ruins communicate through wireless communication signals. Through the information exchange between the search and rescue robot inside the ruins and the control station system outside the ruins, the control station obtains the information inside the ruins and the robot's hardware and software operation data, and the search and rescue robot obtains the control instructions from the control station. Since wireless communication signals are essentially electromagnetic signals, the propagation of electromagnetic signals through obstacles within the building rubble can lead to severe attenuation of the communication signal, greatly reducing the scope of the robot's search and rescue operations. When controlling the robot to climb the stairs and search for survivors on different floors, the staff can only operate the robot through the video signal and distance data detected by the distance sensor inside the ruins transmitted by the robot. Directly lead to the reduction of search and rescue efficiency, and the efficiency of search and rescue work is extremely important in the extremely valuable time of the ruins search and rescue work. Moreover, with the widespread use of various sensors in search and rescue robots, there are measurement errors in the sensor measurements in bumpy environments, and the existence of measurement errors directly leads to decision-making errors in robot control. And communication interruptions can cause the robot to lose contact and lose control within the rubble.

Research on the hardware system, control system, control station system, and autonomous movement of the ruin search and rescue robot will provide a comprehensive understanding of the ruin search and rescue robot. On the one hand, while the successful application of in-depth research results in related aspects will expand the stability and efficiency of the robot, which will further enhance the efficiency and rescue effectiveness of disaster debris search and rescue work.

6.1.2 Research Trends in Ruin Search and Rescue Robots

Although the research of search and rescue robots started late, the research of robots has been in-depth for many years and has a wealth of referenceable research results. According to the specificity of the working environment of search and rescue robots and the specificity of the work content, the research of search and rescue robots is more focused on the improvement of environmental adaptability. According to the development trend of mobile robots and the special needs of search and rescue work, the development trend of search and rescue robots can be summarized in the following aspects.

- Study of stable and reliable intra-ruin communication methods

At present, the communication between the search and rescue robot and the operator is divided into two ways: cable communication and wireless communication. Cable communication is stable and reliable, but the scope of search and rescue operation of the robot is seriously hindered in the ruin environment; wireless communication is convenient and flexible, but the obstacles within the ruins can lead to serious attenuation of communication signals, and at the same time, various electromagnetic interference also seriously affects the stability and reliability of wireless communication. A stable and reliable communication method inside the ruins is one of the urgent problems to be solved.

- Research on autonomous motion control method in ruined environment

Due to the harsh conditions in the debris environment, the autonomous motion control method of structured and level ground mobile robots can hardly meet the demand of autonomous robot motion in the debris environment, while the autonomous motion of search and rescue robots in the debris environment can not only effectively improve the efficiency of search and rescue work, but also effectively expand the scope of search and rescue work of robots. The study of autonomous robot motion control methods for the debris environment is imperative.

- Multi-robot search and rescue team research

Different structures of search and rescue robots have different movement capabilities, and different functions of search and rescue robots can perform different rescue tasks. The complexity of the ruin environment and the diversity of rescue tasks require a variety of different search and rescue robots to participate in the rescue work, and each robot coordinates with each other to quickly complete the search and rescue tasks.

6.2 Hardware Systems of Ruin Search and Rescue Robots

6.2.1 Hardware Composition of the Ruin Search and Rescue Robot

The hardware system of the ruin search and rescue robot consists of the robot body hardware, motion actuator hardware, sensing system hardware and various hardware systems of life detection equipment.

The body hardware mechanism of the ruin search and rescue robot, although its movement is divided into different movement methods such as wheeled and tracked, the robot's body hardware structure is similar; the robot's movement actuator contains electric and pneumatic methods; and the sensing system contains distance sensing equipment, sound sensing equipment, image sensing equipment and other sensing equipment; the various types of life detection equipment carried by the robot are different due to their working The basic principle of the robot is different, and the hardware composition is also very different.

Here, the successful application of a typical debris search and rescue robot, the Deformable Search and Rescue robot System, is used to illustrate the typical system components of a debris search and rescue robot, and the kinematic model of the robot.

6.2.2 Hardware System of Deformable Search and Rescue Robot

The deformable debris search and rescue robot developed by the State Key Laboratory of robotics, Shenyang Institute of Automation, Chinese Academy of Sciences is used as an example to study and explain the hardware system and control system of the debris search and rescue robot. As a disaster rescue robot, the deformable debris search and rescue robot can be equipped with cameras, pickups and other equipment to obtain environmental information inside the debris, and can also be equipped with professional life detectors to detect the signs of life of survivors inside the debris. The deformable debris search and rescue robot is an extension of human legs and feet, eyes and ears and other organs, and plays a huge role in the process of debris search and rescue.

- Deformable ruin search and rescue (SAR) robot system hardware composition and characteristics

The deformable SAR robot is a highly mobile robot with a changeable configuration that adapts to different operating conditions by changing its own configuration. The modular structure of the deformable SAR robot is not only conducive to maintenance

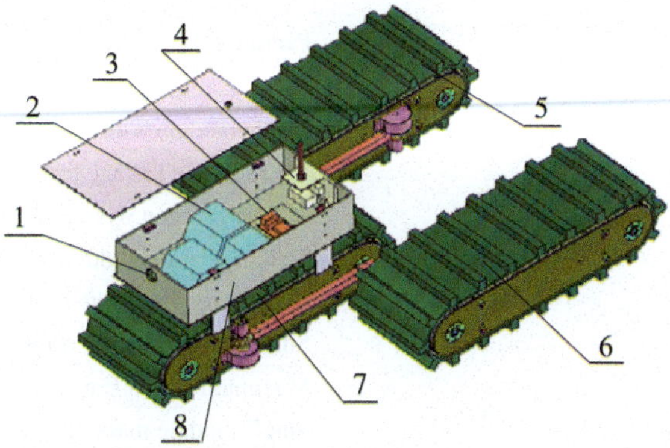

Fig. 6.1 Overall structure of deformable search and rescue robot. 1. Camera; 2. Power supply; 3. Main control unit; 4. Wireless communication module; 5. Module C; 6. Module A; 7. Module B; 8. Head

in a tense rubble search and rescue site, but also to mass production. Figure 6.1 shows the overall structure of the deformable SAR robot.

The motion body of the robot consists of three independent modules, Module A, Module B and Module C. Among them, Module A contains a crawler drive and a pitch drive with two degree of freedom, Module B contains a crawler drive, a pitch drive and a module deflection device with three degree of freedom, and Module C contains a crawler drive and a module deflection device with two degree of freedom. The mechanical links between module A and module B, and between module B and module C are made by detachable connecting rods. The power supply, main control unit, wireless communication module and camera of the robot are distributed in the head above the robot module B.

The parameters of the deformable search and rescue robot are shown in Table 6.1, where the d-configuration, T-configuration and L-configuration are the three typical configurations of deformable search and rescue robots, respectively.

The deformable search and rescue robot has the following characteristics.

- Modular structure design: easy to assemble and maintain, easy to reconfigure.
- Crawler drive: capable of adapting to non-flat, obstacle-ridden ruin environments with high mobility.
- Reconfigurable configuration: three modules are mechanically linked by connecting rods and can form various configurations by changing the topology between modules.
- Module internal hollow: Adopting the design concept of external package and internal hollow, the internal is electrical wiring and the external performance is track drive, which has good protection ability.
- Movement and piggybacking: High mobility ensures good robot movement, and can piggyback all kinds of sensing equipment and search and rescue equipment, which is conducive to expanding both robot performance and operational capabilities.

• Deformable debris search and rescue robot motion mechanism

Table 6.1 Parameters of the deformable search and rescue robot

Parameters	Numerical value
Ground length per module (mm)	276
Single module width (mm)	110
Inter-module linkage length (mm)	193
D-configuration width (mm)	380
T-configuration width (mm)	540
L-configuration width (mm)	280
Single module height (mm)	150
Maximum height (mm)	240
Single module mass (kg)	3
Overall mass (kg)	20
Motor rated power (W)	10
Motor rated voltage (V)	24
Motor maximum torque ($N \cdot m$)	10
Maximum speed of crawler drive (m/s)	1.3
Maximum acceleration of track drive (m/s^2)	1

The motion of the deformable SAR robot can be mainly divided into linear motion, steering motion, pitch motion and conformal transformation, and various complex motions are constituted by the superposition of motion combinations between the robot modules.

- Linear motion of the robot

The linear motion of the robot consists of a combination of linear motion superimposed by three modules driven by the crawler drive module, as shown in Fig. 6.2a, which shows the force analysis diagram of the upper point of the crawler device during the linear motion of the single module of the robot driven by the crawler drive. In this case, the drive motor J_i moves clockwise to drive the crawler motion. The forces on a point P_i on the track are the pressure N_1 of the drive motor, the elastic force N_2 of the ground, the pressure N_3 exerted by other parts of the robot, the force F_1 of other parts of the track on the point, and the frictional force f of the ground. in the vertical direction $N_2 = N_3$; in the horizontal direction when the robot does not slip sideways with the ground $f = F_1 + N_1$. The individual modules of the robot perform linear motion driven by the track drive. As shown in Fig. 6.2b, V_A, V_B and V_C are the linear motion speeds of module A, module B, and module C, respectively, $V_A = V_B = V_C$ when the robot performs linear motion.

- Steering motion of the robot

The steering motion of the deformable search and rescue robot is done by the differential speed method, which is accomplished by the cooperative motion movement between different modules. As shown in Fig. 6.3a, the motion speed of each module

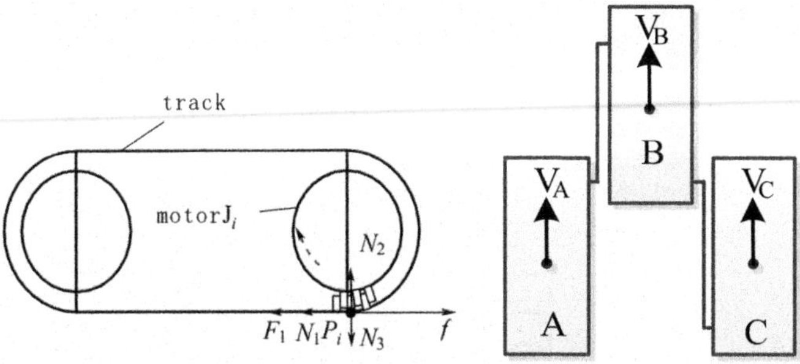

(a) (b)

Fig. 6.2 Schematic diagram of linear motion

of the robot during the steering motion of the robot is shown schematically. When the robot is in steering motion, the motion speed of the B module of the robot $V_B = 0$, the linear motion speed of module A and module C are in opposite directions and the same amplitude $|V_A| = |V_C|$. Since the three modules are connected by rigid bodies, the robot does not deform when module A and module C move at different speeds, and sliding friction occurs between the robot and the ground. Figure 6.3b shows the force analysis diagram of each module during the steering motion of the robot, where f_A, f_B and f_C are the sliding friction between module A, module B and module C of the robot and the ground, and r_A, r_B and r_C are the straight line distances from the center O of the robot to f_A, f_B, f_C respectively.

Due to the sliding friction between the robot and the ground, the robot generates a steering motion under the moment of the sliding friction of each module $M = r_A \times f_A - r_B \times f_B + r_C \times f_C$.

- Mechanism of robot deformation

Module A and module C of the deformable robot have two degrees of freedom, and module B has three degrees of freedom. The relative positions of the three modules can be changed by the pitching device and the deflection device of the robot, and then the configuration of the robot can be changed. As shown in Fig. 6.4a, J_2 and J_4 are the

Fig. 6.3 Schematic diagram of steering motion

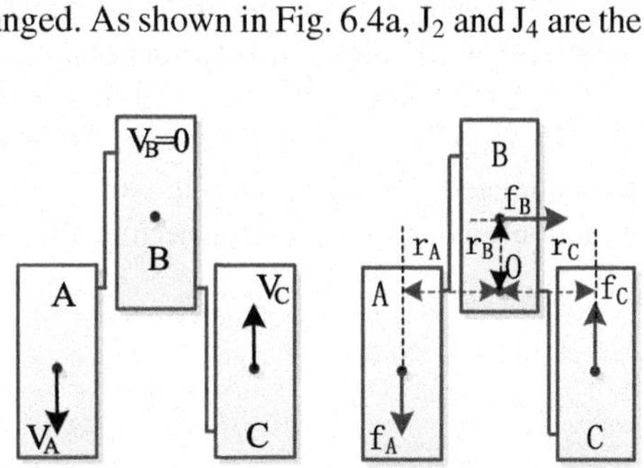

(a) Schematic 1 (b) Schematic 2

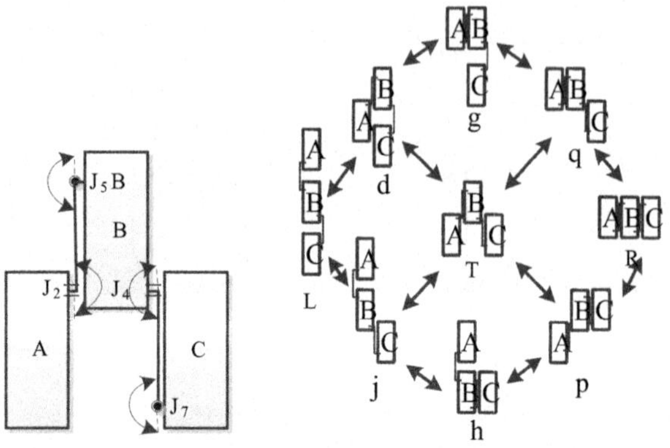

(a) Diagram 1 (b) Diagram 2

Fig. 6.4 Deformable robot deformation mechanism diagram

pitching device of the deformable robot module A and module B, respectively, can be driven to make the modules A and B of the robot pitch in the vertical direction, J_5 and J_7 the deflection device of the deformable robot module B and module C, respectively, can be driven to make the modules B and C of the robot deflect in the horizontal direction, respectively. As shown in Fig. 6.4b, by driving the pitch and deflection devices of the robot, the deformable robot has 9 different conformations, namely, g conformation, d conformation, q conformation, L conformation, T conformation, R conformation, j conformation, p conformation, and h conformation, and the arrows along the way mark the transformation paths between the different conformations.

The T configuration has good steering performance, passing performance and anti-rollover stability, and is the most widely used configuration for roaming, anti-rollover and stair climbing in the search and rescue of debris. The D configuration has better passing performance than the T configuration, and also has steering and overturning performance, but its anti-rollover stability is weaker than that of the T configuration, and this configuration is usually used for traversing narrow rubble spaces; the L configuration has the best passing performance, but the steering performance and anti-rollover stability of the robot in this configuration are not good, and this configuration is mainly used for traversing narrow spaces in relatively gentle environments; the R configuration has the best passing performance, but the steering performance and anti-rollover stability of the robot in this configuration are poor. The T configuration is the best overall performance motion configuration, therefore, the initialized configuration of the robot is the T configuration, and it is also the most widely used configuration for performing rubble search and rescue tasks.

6.2.3 *Kinematic Model of Deformable Search and Rescue Robot*

The kinematic model of a robot is the basis for robot motion analysis and development of robot motion control strategies. Since the deformable search and rescue robot performs the actions of roaming, crossing obstacles and climbing stairs inside the ruins mostly based on the T-configuration, which is also the best comprehensive performance configuration of the deformable robot, the kinematic model of the deformable robot in this paper is the kinematic model of the robot in the T-configuration.

The deformable robot is a tracked robot. When the robot performs linear motion on a plane, the track drives of the three modules drive at the same speed, i.e., $\omega_A = \omega_B = \omega_C$ the robot moves at

$$V = \omega_B * r_m \tag{6.1}$$

where

- ω_A angular velocity of the track drive of the deformable robot module A, rad/s.
- ω_B angular velocity of the track drive of the deformable robot module B, rad/s.
- ω_C angular velocity of the track drive of the deformable robot module C, rad/s.
- r_m distance from the robot track drive to the track surface, m.

Figure 6.5a shows the side view of the robot in the "head-up" attitude for linear motion. When the deformable robot performs steering motion, according to the analysis above, the deformable robot uses the differential speed method for steering, i.e., module A and module C move with opposite directions and the same magnitude, as shown in Fig. 6.5b, which is the top view of the steering motion of the deformable robot. Where, V_A and V_C are the motion velocities of single module linear motion of deformable robot module A and module C with $V_A = \omega_A{}^*r_m$, $V_B = \omega_B{}^*r_m$ respectively. The steering motion of the deformable robot is realized by the cooperative motion of module A and module C. The steering angular velocity of the robot is

$$\omega_\theta = \frac{V_C'}{r_\theta} \tag{6.2}$$

The steering angle of the robot is

$$\theta = \omega_\theta{}^*t \tag{6.3}$$

where V_C'—the actual speed of movement of the center of gravity of the deformable robot module C, m/s.

Since the modules A and C of the deformable robot experience sliding friction with the ground during the steering motion, therefore $V_C' \neq V_C$; t is the motion time of the steering motion of the deformable robot.

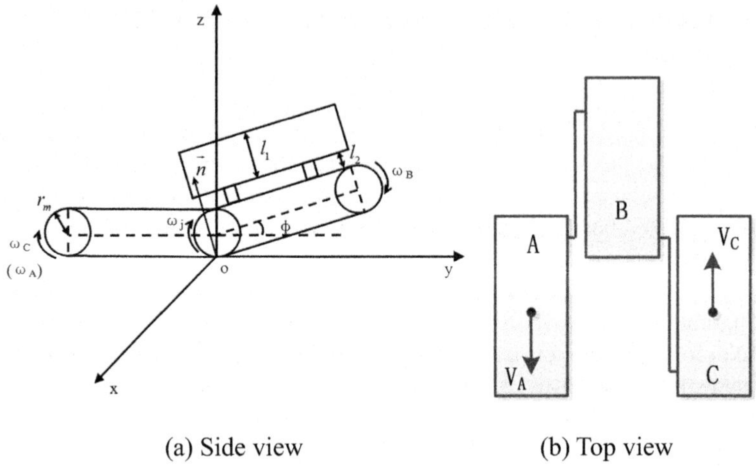

Fig. 6.5 Schematic diagram of deformable robot

The deformable robot uses the tilt angle of the robot head to describe the tilt angle of the deformable robot, and the direction vector $\vec{n}$ perpendicular to the robot head to describe the tilt degree of the deformable robot. When the robot is not tilted:

$$\vec{n} = \vec{n_0} = \begin{bmatrix} 0 \\ 0 \\ 1 \end{bmatrix} \tag{6.4}$$

When the deformable robot tilts θ_y in the left and right directions, which is equivalent to the deformable robot rolling sideways θ_y, the robot's tilt direction vector is

$$\vec{n} = R(y, \theta_y)\vec{n_0} = \begin{bmatrix} \cos(\theta_y) & 0 \sin(\theta_y) \\ 0 & 1\ 0 \\ -\sin(\theta_y) & 0 \cos(\theta_y) \end{bmatrix} \begin{bmatrix} 0 \\ 0 \\ 1 \end{bmatrix} = \begin{bmatrix} \sin(\theta_y) \\ 0 \\ \cos(\theta_y) \end{bmatrix} \tag{6.5}$$

When the deformable robot tilts θ_x in the left and right directions, which is equivalent to the deformable robot pitching θ_x, the robot's tilt direction vector is

$$\vec{n} = R(x, \theta_x)\vec{n_0} = \begin{bmatrix} \cos(\theta_x) & 0 \sin(\theta_x) \\ 0 & 1\ 0 \\ -\sin(\theta_x) & 0 \cos(\theta_x) \end{bmatrix} \begin{bmatrix} 0 \\ 0 \\ 1 \end{bmatrix} = \begin{bmatrix} \sin(\theta_x) \\ 0 \\ \cos(\theta_x) \end{bmatrix} \tag{6.6}$$

When the deformable robot is climbing over obstacles and stairs, the angular velocities of the drive motors of the track drive unit and the drive motors of the pitch unit of each module of the robot need to be coordinated and controlled in order to adjust the robot's posture. The angular velocities of the drive motors of each module of the deformable robot show a certain proportional relationship so that the robot can adjust the robot's position and attitude through a series of motion movements. The proportional relationship between the angular velocities of the drive motors of each module drive unit and the drive motors of the pitch unit of the deformable robot is as follows.

$$S_A = \frac{\omega_A}{\omega_B} \tag{6.7}$$

$$S_C = \frac{\omega_C}{\omega_B} \tag{6.8}$$

$$S_j = \frac{\omega_j}{\omega_B} \tag{6.9}$$

where

ω_A angular velocity of the track drive of the deformable robot module A, rad/s.

ω_B angular velocity of the track drive of the deformable robot module B, rad/s.

ω_C angular velocity of the track drive of the deformable robot module C, rad/s.

ω_j angular speed of the drive motors of the pitching devices of the deformable robots module A and module B, rad/s.

S_A, S_C and S_j is the ratio of the angular speeds of the different motors.

6.3 Control System of Ruin Search and Rescue Robot

Ruin search and rescue robots are characterized by both high hardware complexity and high control complexity, as well as the need to deal with the unstructured external environment, therefore, the control system of ruin search and rescue robots has higher requirements in terms of reliability and stability compared with traditional robot systems.

6.3.1 Requirements for the Control System of Ruin Search and Rescue Robot

In order to achieve flexible manipulation of the robot and further realize the autonomous motion of the robot, the deformation ability and diverse motion postures of the robot increase the difficulty of the control system design. The control system of this type of robot needs to meet the following requirements.

- Real-time

This type of robot is oriented to the operation environment of complex terrain, so the first step is to realize the flexible control of this robot, and the real-time of the control system is the basic requirement. However, the control system has to perform many tasks at the same time, such as planning the motion of each joint, filtering and identifying the feedback signals from each sensor, feeding the operator with its own operating status and some necessary information, and handling any malfunctions in a timely manner. In order to make the motion of the joints smooth and responsive, or at least to make the operator feel less stuck, the control cycle of the operating system for the joint motors should be less than 200 ms, and the robot has five moving joints that need to be controlled simultaneously. For sensor signal acquisition, if the application of the information is not considered in terms of cost, the higher the frequency of acquisition, the better, and the higher the credibility of the final information obtained by processing a large amount of data. Also communication takes time. It can be seen that the requirement of real-time is necessary and difficult for this type of robot.

- Operational Integrity

The robot has a variety of amphibious gait, and can be deformed, with strong environmental adaptability. The operator should be able to control the robot in any configuration and gait, taking advantage of the robot's mechanical design as much as possible. The control system builds a bridge between the operator and the robot, and should provide the operator with the ability to manipulate the robot at will, while protecting against actions that could damage the robot and avoid operator misuse.

- Fault tolerance

The control system should be able to control the robot's motion status and the operation status of each electronic device at all times, and be able to take protective measures automatically. For example, at the beginning of power-up, the control system should perform a self-test of the electronics. The robot should also be tested regularly during its motion, and if a fault or possible damage to the robot is detected, it should be able to stop the motion automatically or return to a stable attitude.

- Scalability

Robots face a variety of tasks and sometimes need to add sensors, additional control strategies, etc. depending on the task. The control system of a robot should be easily expandable without changing the overall system framework, altering or removing functions that have already been implemented.

- Monitoring the robot operation status

The operating status of the robot needs to be monitored in real time and the data saved for post-processing, commissioning and maintenance.

6.3.2 Structure Design of Hierarchical Distributed Modular Control System

To meet the requirements described in the previous section, a modular control system structure was designed based on a hierarchical system and step-by-step control, as shown in Fig. 6.6. The control system is divided into three layers: the monitoring layer, the planning layer, and the execution layer. The monitoring layer implements human–machine interaction, the planning layer plans the control strategy, and the execution layer implements motion control and sensor signal acquisition and pre-processing. Each layer consists of one or more functional modules with independent controllers, which are connected by a multi-master bus communication to form a distributed control system.

The monitoring layer implements human–robot interaction, with a friendly interface for the operator to operate the robot and feedback to the operator on the robot status. The monitoring layer consists of a monitoring platform and a wireless communication module. The monitoring platform sends operation commands, displays and stores the robot status information; the wireless communication module realizes the communication between the monitoring layer and the planning layer.

The planning layer carries out control strategy planning. The planning layer connects the monitoring layer with the execution layer and is at the center of the entire control system, playing the role of a scheduling center. This layer performs global planning of the robot's motion based on operator commands, robot status

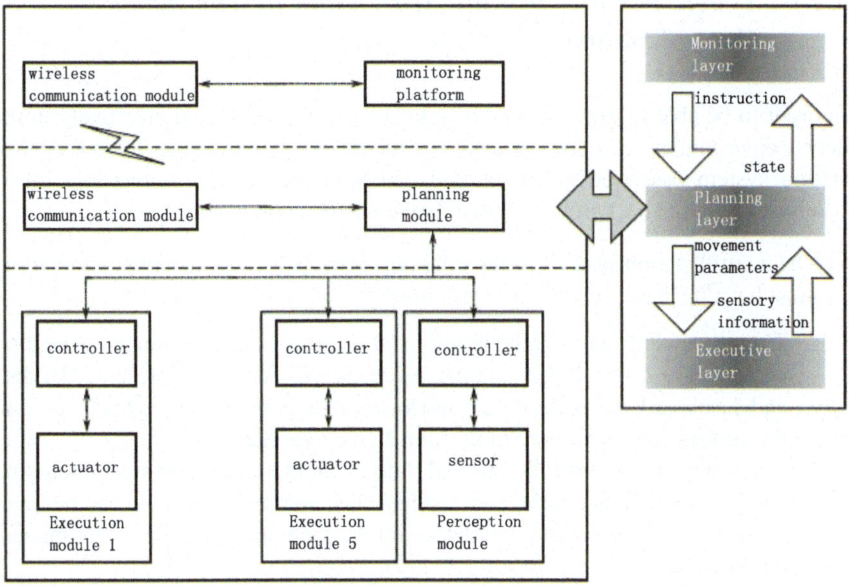

Fig. 6.6 Control system structure block diagram

information, environmental information, and specific tasks. This layer consists of a planning module and a wireless communication module.

The actuation layer implements motion control and sensor signal acquisition and pre-processing. The actuation layer consists of multiple actuation modules and multiple sensing modules, with the largest number of functional modules. The actuator modules are used to control the motion of individual joint motors, reducing the control cycle to less than 100 ms; the sensing modules are used to collect sensor information and perform pre-processing such as software filtering and recognition.

6.4 Ruin Search and Rescue Robot Control Station System

Search and rescue robot system is a typical "human–machine-environment" system, including not only the robot body mechanism, but also the sensing, communication and control station and other ontological support systems. In the search and rescue operation, the robot and the operator are usually in the remote operation state of man–machine separation, and the operator can only obtain the environment and the state of the robot through the control station, and plan and control its execution of rescue tasks. Therefore, the control station is one of the core control mechanisms of the robot system and is the only platform to realize human–robot interaction.

6.4.1 System Features of Ruin Search and Rescue Robot Control Station

In order to be able to perform search and rescue work safely and effectively in the debris environment, this section analyzes the design requirements for the control station system based on the functional characteristics of the deformable robot, combined with the background of search and rescue applications.

- The control station system, as the only platform for human–machine interaction, must have the basic functions of monitoring.
  - The robot status, environment and location, and control command execution status can be displayed in real time. The robot status can be represented by data and dashboard, the environment and location by video and audio feedback, and the control command execution status by text description.
  - The robot can be controlled in real time to change its motion state, including speed, direction and parameter settings. The control commands can be input by soft keys, buttons and joysticks, and the parameters can be set by manual modification.
- Considering the special application context oriented to search and rescue operations, the design should also meet the following requirements.

- The survival capability of search and rescue robots depends mainly on the overall environmental adaptability of the system. The disaster-stricken environment has potential hazards such as secondary collapse, and the environment in which the control station is located may change. The system should achieve composite functions and reduce weight, volume, and power consumption as much as possible. While adapting to the environment, it should cope with sudden external disturbances as much as possible.
- The high-hazard nature of the disaster environment poses a serious threat to the safety of the robot itself, and the high complexity of rescue operations can easily cause physical and cognitive fatigue for the operator, and the robot can sometimes develop unexpected conditions. The control station should provide solutions to safety hazards such as communication delays, mechanism failures, non-structural dynamic environments and misuse whenever possible.
- The basic idea of search and rescue robots is to realize human-environment interaction through control stations, which combine human perception and behavioral capabilities extends to the remote hazardous environment. The control station design should take into account physical barriers, signal attenuation, and other interference factors, and can integrate visual and non-visual information using multiple effect channels. Provide as many views as possible to observe the robot's position information within the local environment, in addition to the proprioceptive observation angle.
- The mobile needs of the search and rescue robot and the relatively fixed location of the control station operations put the system in a state of remote operation with human-machine separation. This state seriously increases the complexity of the rescue. The control station should have local autonomous capability to ease the work intensity of the operator. Rescue operations on the control station monitoring accuracy, interaction quality, portability and comfort have high requirements, the design should be as friendly and humane as possible.
- In practice, most of the applications use ground rescue robots, airborne unmanned helicopters, ground rescue staff, medical personnel and ground rescue vehicles together to form a multi-intelligent body of three-dimensional network system. Control station system should be able to give full play to the characteristics and advantages of each platform, auxiliary interactive information flow and sharing in the overall scope of the system, providing real-time accurate information and action support for the judgment and decision-making of rescue units at all levels.

• The following specific functional requirements should also be taken into account during the design of the control station.

  - Deformable robots have strong environmental adaptability and high mobility, and multiple motion configurations and gait patterns make them have different environmental and spatial adaptability. Considering the different complexity of control tasks, the control station should provide multi-level selectable control modes to meet the needs of deformable robots with many control commands.

  - Deformable robots have certain autonomous capabilities, and when the operator cannot make accurate judgment of the environment of the robot is in time, the control station system should provide a call interface for the robot's autonomous steering, obstacle crossing and emergency stop functions to realize a cooperative work mode combining manipulation and supervision.
  - The robot control station system connects the robot and the operator in a closed-loop loop from the information level, and manipulates the robot to complete search and rescue tasks in the ruins and other environments through environmental perception, auxiliary operation, information feedback and other interactive methods, realizing the extension of human perception and action capabilities.

According to the functional requirements of search and rescue robots analyzed in the previous section, a control station system based on human–computer interaction technology should follow the following principles.

- Environmental adaptability: Search and rescue sites are usually non-structural and complex environments caused by random disasters, and the environment in which the operator is located may change with different needs, while the mission deployment will also change according to the specific disaster conditions. Therefore, the control station system should have the ability to cope with external instability and uncertainty of anti-interference, with a wide range of environmental adaptability.
- Safety and stability: Safety and stability are the basic prerequisites for the overall function of the system, and the safety of the surrounding people and environment as well as the safety of the robot itself should be considered. The complex and dangerous environment of the search and rescue site and the operator's restricted field of view can lead to a decrease in perception and inference, and the robot can sometimes fall over or get stuck unexpectedly. The control station should provide timely alarms and emergency treatment in case of sudden changes in sensor data to avoid deterioration and ensure smooth execution of the mission.
- Intelligent professionalism: The control station system should have the ability to synergistically combine the flexibility and adaptability of a human with the accuracy and rapidity of a robot. When the robot has difficulty in accurately predicting collisions and selecting strategies in real time, the operator needs to assist the robot in quickly extracting useful information from cumbersome and redundant data and making reasonable inferences. At the same time, the operator needs the robot to provide a basis for decision making through advantages such as rapid and accurate computation and information synthesis. When it is difficult for the operator to make accurate judgments and decisions in a timely manner, the control station should have an auxiliary control planning capability to solve the difficulties caused by the randomness of non-structural environmental obstacles for robot motion planning and to reduce the time consumption of information processing and inference.
- Multi-channel information fusion: the rescue site environment is complex and dangerous, a single feedback information is difficult to meet the actual search

and rescue needs. For example, simple video image information has a limited field of view and is susceptible to dust and smoke, while simple sound information is difficult to identify the source orientation and susceptible to on-site noise interference. Therefore, based on the multi-resource theory, the complementary fusion of parallel multi-channel information resources can enhance the accuracy of interactive information and improve the fault tolerance and robustness of the system.

– Friendly and humane: the control station should reduce the technical training requirements for the operator as much as possible, and enable the operator to perform tasks relatively intuitively and effectively with the aid of the robot. Search and rescue robot operations lack a sense of presence and interactivity, the control station should provide three-dimensional presence and multi-effect channel experience, in line with the three-dimensional intuitive field of activity, non-precision decision-making and implied interaction and other human daily life habits.

6.4.2 System Structure of Ruin Search and Rescue Robot Control Station

With the continuous development of technology level, the study of robot control architecture has gradually transitioned from single hardware, independent professional controller and individual control to software and hardware integration, general open and multi-level coordinated control.

The objective of the robot control station system architecture research is to design a general-purpose robot control station structure with open and modularized architecture. In terms of hardware, an open general-purpose computer platform is used, and its mature hardware and software resources are used as technical support for the expansion of the main controller functions; in terms of software, a standard operating system is used, combined with a highly portable language to write applications.

The division of functional modules, the information interaction pattern between modules and the implementation of functional modules are the focus of research on the architecture of robot control systems. There are usually two basic structures based on hardware hierarchy and functional division, among which, the hardware hierarchy-based division type is relatively easy, but has the defect of strong dependence on hardware. The function-based architecture analyzes and considers the hardware and software system as a whole from a functional point of view, and this type of division is more in line with the original research intention of the architecture.

According to the different ways of connection between the components established by the architecture, it can be divided into three types of architecture: deliberate, behavior-based, and hybrid.

– Deliberative architecture: is a robot's reasoning to generate action instructions for a desired goal based on known logical knowledge or search methods. Knowledge representation of resources, tasks, and action goals is accomplished through a logical language and generation rules. Explicit with strong reasoning ability and knowledge representation, mainly for applications in structured environments.

- Behavior-based architecture: It refers to the decomposition of the system into basic behaviors with their own control mechanisms, and the ability to adapt and respond to changes in the external environment through the combination of corresponding arbitration mechanisms to form intelligent actions. To a certain extent, it alleviates or even avoids the work of designing and building environment models.
- Hybrid architecture: It is a combination of deliberative and behavior-based architecture integration mechanisms, using a master actuator as a sequencer to acquire actions, make decisions and execute tasks through the planning layer. It usually includes five basic components: sequencer, resource manager, mapper, task planner and performance supervised solver.

The design and establishment of the robot architecture includes several design points that can be drawn from the distributed nature of computing, independence of communication, flexibility of the system, scalability, and remote supervision and control.

According to the analyzed robot control requirements and SAR operation requirements, based on the proposed five design principles of environmental adaptability, safety and stability, intelligent professionalism, multi-channel information fusion and friendly humanization, and combined with the task characteristics and working environment of SAR robots, the control station system architecture is established using a hybrid architecture integrating deliberate and behavior-based architecture, as shown in Fig. 6.7. The following is divided into four aspects: hierarchical division, control loop, functional module and data flow for analysis and explanation.

- Hierarchy of architecture

The architecture consists of Supervision and Coordination Level, Control and Planning Level, and Executive and Feedback Level. It can meet the environmental adaptability requirements of debris search and rescue robots, and categorize a wide range of functional requirements of deformable robots in a hierarchical design, which improves the versatility and reliability of the system.

- Executive and Feedback Level

The execution feedback layer is the lowest layer where the robot system interacts directly with the external environment. Through the sensing feedback information processing module and the environmental information and robot state sensing module, the data collection and information processing of environmental parameters, robot position, task execution state, etc. are realized; through the robot motor drive control module combined with the first module, intermediate module and tail module respective motor controllers execute the upper layer control commands based on the current state, such as steering, speed regulation, deformation and other control commands.

- Control and Planning Level

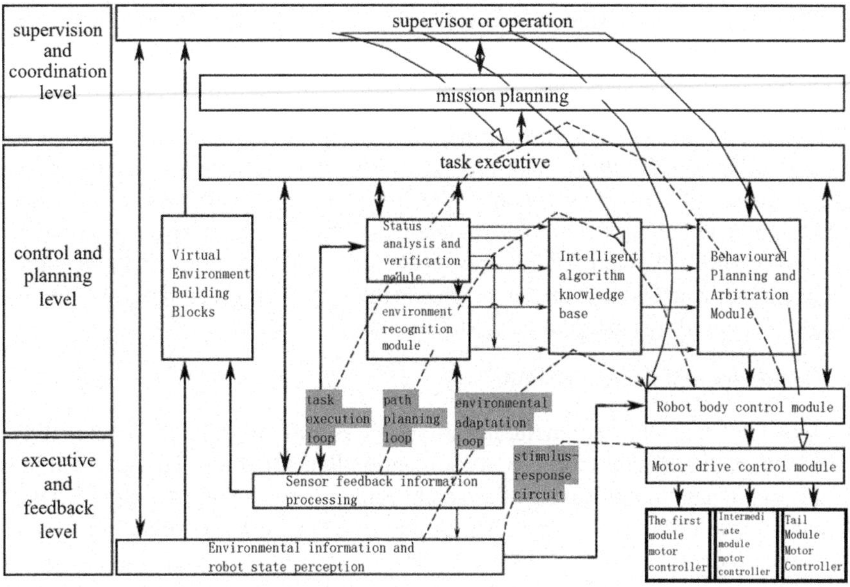

Fig. 6.7 Control station system architecture diagram

The control planning layer includes the virtual environment construction module, the state resolution inspection module, the environment recognition module, the intelligent algorithm knowledge base, the behavior planning and arbitration module, and the robot vehicle control module. This layer is an intermediate layer between the top-level supervision and coordination layer and the robottom-level execution and feedback layer, and is responsible for the interaction between the two layers.

First, the task decision is executed according to the manipulation or supervision mode selected by the supervision and coordination layer, and the task execution module is responsible for scheduling and planning for each other functional module, which is the core part of control planning at this level. In the manipulation mode, the task execution is directly assigned to the robot body control module; in the supervisory mode, the task execution needs to pass through the behavior planning and arbitration module of the control station system and then execute the control of the robot body.

Secondly, the environment and robot state information obtained from the robottom execution feedback layer is transmitted to the task execution module either through the environment identification module and the state analysis verification module or directly to the task execution module as a basis for real-time determination of the task execution state and decision making for control planning. The virtual environment construction module fuses the state information to construct a 3D virtual environment of the robot in the local environment, and the environment and robot perception information is transmitted to the upper supervision and coordination layer either through this module or directly.

– Supervision and Coordination Level

The supervisory planning layer is the highest level of intelligence in the architecture and consists of a control/supervisory module and a transactional decision module. The operator supervises, controls, and makes control decisions about the robot's disaster relief operations through this layer. The operator's involvement improves the overall intelligence of the robot system, including global environment awareness, rapid implementation of robot task planning and transaction decision making, and the choice of roboth manipulation and supervision modes ensures the application of robot autonomy and solves the unknown, complex, dangerous, and high-load rescue challenges posed by unstructured environments.

- Control loop of the architecture

Four parallel control loops, including stimulus–response loop, environment adaptation loop, path planning loop and task execution loop, are depicted as dashed lines in the bottom and middle layers of the architecture, reflecting the autonomous intelligence of the deformable robot. The stimulus–response loop realizes emergency stop in case of sudden change of robot state; the environment adaptation loop realizes collision detection and autonomous obstacle avoidance during robot navigation; the path planning loop is responsible for robot environment recognition and path tracking; the task execution loop realizes local autonomous behavior according to the knowledge base of decision scheduling intelligence algorithm in the supervisory planning layer.

The four dotted lines running through the three levels of the architecture depict the operator's ability to intervene and control the three control loops of environmental adaptation loop, path planning loop and task execution loop through the control station system. At the same time, the operator can directly control the motor drive control module to achieve basic bottom-level control or debugging functions such as steering, reflecting the environmental adaptability and intelligent professionalism of the control station system.

- Functional modules of the architecture

In order to describe the functional modules and information interaction process more clearly, this paper simplifies the proposed control station system architecture diagram, focusing on the functions of each module and information interaction, and gives a block diagram of the functional structure of the control station system, which is an integrated human–machine-environment robot structure consisting of a robot and a control station system, as shown in Fig. 6.8.

The robot system includes 3 subsystems in addition to the body mechanism.

Main control subsystem: analyzes control commands and state information, plans robot motion configuration and gait, and is the decision-making system of the robot.

Execution subsystem: executes the robot's motion planning, changes the motion configuration and gait.

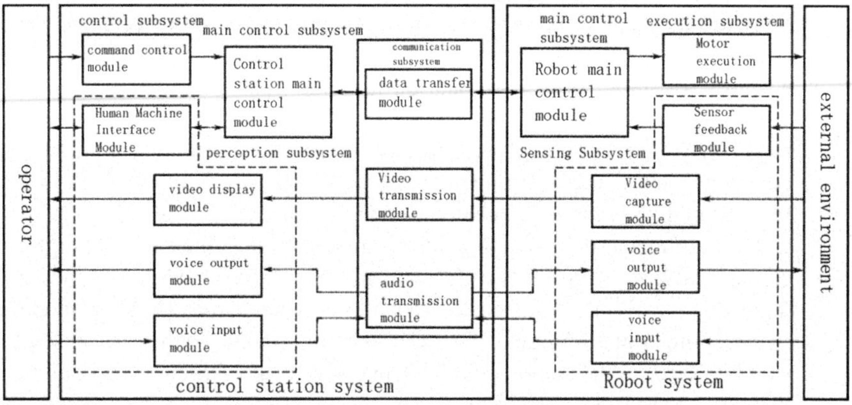

Fig. 6.8 Block diagram of the functional structure of the control station system

Sensing subsystem: capable of detecting and sensing the surrounding environment through multiple sensors, realizing the coexistence of multi-channel sensing information.

The control station system consists of four subsystems.

Master control subsystem: As the core part of the control station system, it is responsible for the integration of information and planning decisions, and manipulates the robot to execute planning deployment. The main control module of the control station extracts, decodes, analyzes, optimizes, coordinates and stores the interaction information, and provides a human–machine interface for the closed-loop interaction between the operator and the robot.

Manipulation subsystem: Combined with the human–machine interface module and command manipulation module, the subsystem provides a variety of control command input modes based on multiple interactive resources. Operators combined with specific rescue tasks and on-site disaster conditions to switch the control or supervision work mode.

Sensing subsystem: The virtual and real environment is fused through multiple information channels of data, video and audio to provide a basis for the operator's control decisions. It has the functions of environment perception and state perception. Environmental perception provides both auditory and visual channels to obtain information about the external environment. State perception includes robot configuration gait and task execution state perception.

Communication subsystem: As a bridge for information interaction between the control station and the robot, it realizes the accurate issuance of control planning instructions from the operator to the robot control system, and obtains the robot's feedback position and status and rescue environment information in real time to ensure the real-time monitoring and long-distance control of the robot status by the control station system.

- Data flow of the architecture

- Sensing state data: The control station system provides a combination of virtual and real real-time state feedback mechanisms. The control station system acquires real-time robot state data through the transmission subsystem and transmits it to the main control module to build a virtual monitoring environment, which, combined with real environmental information, allows the operator to sense and determine the actual operating state of the robot in real time through the human–machine interface module.
- Control command data: The operator is able to control the robot through the control station system by two mechanisms. First, the supervisory mode and input commands are selected through the command manipulation module. In direct manipulation mode, the master control module of the control station simply processes the commands and delivers the task instructions as command lines; in supervisory control mode, the master control module processes the command tasks to generate a sequence of behaviors, and then implements the command delivery through the data transfer module.

The control station system incorporates a hierarchical control system and a behavior-based control system with clear hierarchy, open structure, and high versatility. At the same time, the modular division by function, with multi-channel information fusion, from the information level of the human (rescue personnel), machine (deformable disaster rescue robot system) and the environment (ruins rescue site) connected in a closed-loop loop, through environmental perception, auxiliary operation, information feedback and other interactive ways to supervise the control robot in the ruins and other environments to perform disaster relief and rescue tasks, to achieve the extend of search and rescue personnel's perception and action ability.

6.4.3 Working Mode of Ruin Search and Rescue Robot Control Station

In the field of robotics research, the human–robot interaction channel is the communication channel that transmits and exchanges information between humans and robots. HCI channels include sensory channels, which are mainly used for sensing information, and effector channels, which are mainly used for processing and task execution based on the sensed information. The development of HCI today tends to capture the user's intention by integrating inputs from multiple channels with different degrees of accuracy, which can reflect the rational computer structure and improve the naturalness and efficiency of interaction.

The visual and voice channels are the natural interaction modes most consistent with human daily habits, and the operator is able to perceive environmental information through the complementary of the two channels. The mechanical contact channel is the most basic interaction mode, which mainly completes the confirmation of task objectives and realizes the issuance of control commands.

The following is an in-depth study of the multi-channel cooperative working mode that the control station system has.

- Visual interaction channel

The narrow collapsed structure of the ruins is difficult for rescuers and even rescue dogs to enter. In order to ensure the safety of field operators, a certain operating distance needs to be maintained between the control station and the robot, i.e., the operator does not follow the robot into the ruins by means of teleoperation, and the research should first solve the video transmission problem of the control station system and the mobile robot body.

At the same time, due to the complexity of the rescue task, the overall SAR charge center needs to unify the supervision and management of the video information detected by each ruin search and rescue robot. Since each robot system has been pre-configured with different transmission systems, the study should also consider how to integrate other robots into the overall SAR robot system without interfering with each other to achieve video information sharing.

According to the above visual interaction channel functional requirements, the control station system uses a variety of video communication modes to work together to build a composite video transmission platform to adapt to different monitoring locations and operational needs, so that the system has functional tailoring, to a certain extent, to reduce the coupling of the system.

The following is a study of composite visual interaction channels for different terminals.

Control station operation site: the visual interaction channel in this part is divided into two real-time state perception channels for real and virtual environments.

The control station system receives the environment and status information collected by the remote robot, and the transmitted real environment information is provided to the human–machine interface module of the control station through the video transmission system. The visual interaction channel of the real environment uses wireless microwave transmission of remote environmental video information, and provides two sets of system design solutions with different carrier frequency and transmitting power for different monitoring locations and search and rescue tasks. They have clear and stable images, better reception quality and strong anti-interference performance, and can adapt to the closed and complex rescue environment features. Provide multiple communication channels to support synchronous transmission of multi-directional audio and video signals collected by the robot, extending the range of vision and realizing the visual perception of the real environment.

Remote command center: In order to adapt to changes in the scale of network monitoring and rescue operation requirements, the design of the visual interaction channel uses a hierarchical network management model, with the remote command center as the network hub of the entire system, each robot control station system as a secondary network node, the control station system and the remote command center visual interaction channel structure schematic diagram shown in Fig. 6.9. Relatively

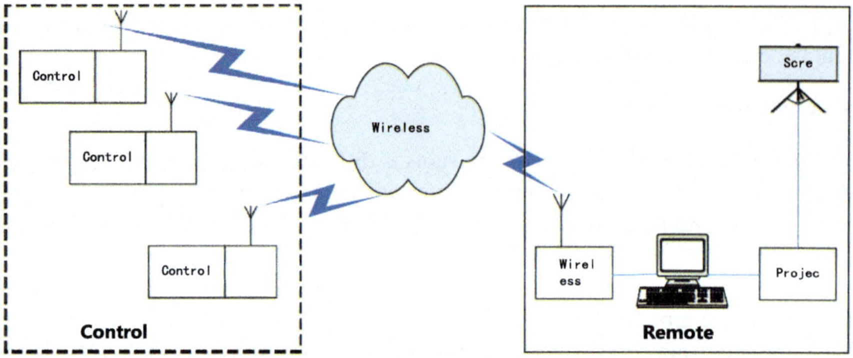

Fig. 6.9 Schematic diagram of the structure of the visual interaction channel between the control operation site and the remote command center

close to the robot rescue operation location of the control station operation site and the remote command center through the wireless LAN to achieve video communication.

Each robot control station placed near different rescue sites receives video monitoring information from their own feedback and transmits video signals via wired means to network video servers, which are connected to the wireless LAN through a wireless router. The PC placed in the remote command center is wired to the public router and connected to the wireless LAN. It is able to access and manage the network video servers in the network through special video surveillance software and decode and display the video information.

- Voice interaction channel

The main goal of the design is to be able to complement the visual channel, alleviate the intensity load caused by a single channel, assist in information acquisition when the video signal is severely disturbed, stay close to natural human habits, and improve the efficiency of searching for survivors.

The station system provides three parts including voice interaction between survivors at the ruin site and control station operators, voice interaction between operators at each control station site, and voice interaction between the remote command center and the control station operation site. The voice interaction channel diagram is shown in Fig. 6.10.

The implementation of the voice interaction channel can be combined with the visual interaction channel, using the same transmission equipment for audio signal transmission, and the collection and acquisition can be achieved through pickups, speakers, microphones and other productized intercom equipment.

- Mechanical tactile channel

Mechanical haptic channels are mainly realized by touching and pressing the mouse and keyboard, etc. Touch panels, joysticks and buttons can be used as auxiliary

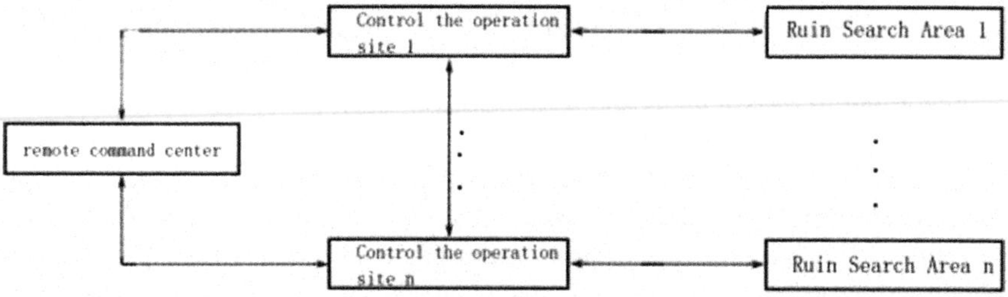

Fig. 6.10 Schematic diagram of the voice interaction channel of the control station system

channels to realize the interaction between human and controller. The mechanical haptic channel interaction process is shown in Fig. 6.11.

In the process of information interaction in the control station system, it is common to combine the characteristics and advantages of different interaction channels to perform interaction tasks in a complementary way with primary and secondary channels. This effectively reduces the load of individual channels and improves the reliability of interaction and task execution efficiency. Figure 6.12 shows the schematic diagram of the cooperative working mode of main and auxiliary channels.

The visual and voice channels of human–computer interaction of the control station system are direct and natural, and are the interaction channels most in line with human daily behavior habits. However, the voice channel needs the guidance and prompting of the visual channel, and the two interaction channels should be used

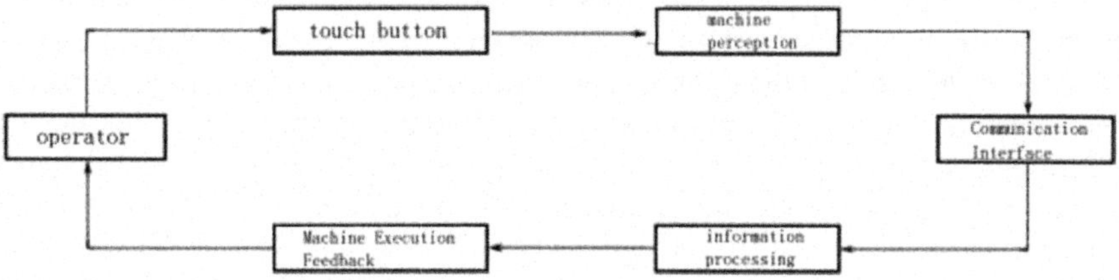

Fig. 6.11 Mechanical haptic channel human–computer interaction process

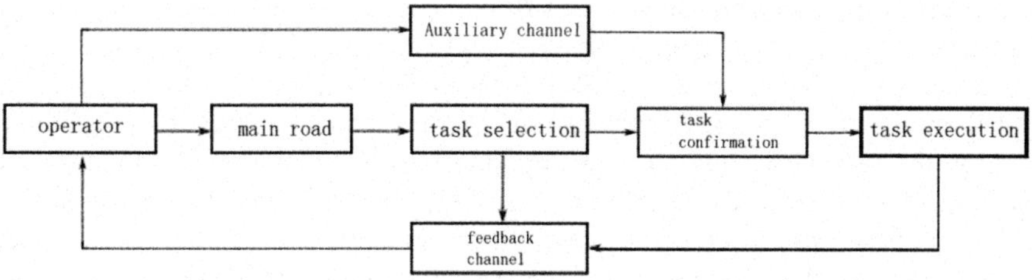

Fig. 6.12 Schematic diagram of the main and auxiliary channels working together

together as the main channel of interaction, while the mechanical haptic channel is used as an auxiliary channel to perform the task decision-making work.

According to the strengths and weaknesses of each effect channel, the primary and secondary synergistic working mode is selected, with the visual combined with audio as the primary channel and the mechanical haptic channel as the secondary channel. The visual and auditory channels take advantage of the fast and efficient perception of information for environmental perception and state acquisition, and when there is a conflict between the two channels, the visual channel information is dominant. The mechanical haptic channel is used as an input means to carry out decision-making tasks and a way to perceive dangerous emergencies.

By working together with multi-channel main and auxiliary, the operating load of a single channel is reduced and operator fatigue is effectively relieved. At the same time, it improves the system perception capability and increases the reliability and efficiency of interaction.

6.5 Autonomous Movement of Ruin Search and Rescue Robots in Bumpy Environment

6.5.1 Effects of Bumpy Environment on Ruin Search and Rescue Robot Motion

The ruin environment is an unstructured and complex environment, and because the ground in the building ruin environment is non-flat, the angle between the plane where the robot is located and the ground plane will change drastically during the robot's movement on the non-flat ground, which is externally expressed as robot bumps. Therefore, the building ruin environment is a bumpy environment. Since the control and decision of the robot's autonomous motion in the building rubble environment depends on the measurement data obtained from the sampling of various sensors of the sensing system, and the various sensors are assembled in a rigid connection, the bumps of the robot cause the angle between the plane where the sensors are located and the ground plane to change drastically, especially for the distance measurement sensors, the measurement direction of the sensors will change drastically, which in turn causes the distance perception data between the robot and the surrounding environment to change drastically. The measurement error of the distance sensing data between the robot and its surroundings increases. In the process of autonomous motion, the distance sensing data is an important basis for motion control and decision making, and the increase of the distance sensing data measurement error will lead to the error of the robot's autonomous motion decision.

Since the robot's decision error is generated by the distance measurement data on which the robot bases its motion with a large error, this paper stipulates that the distance measurement data with a large measurement error is invalid data, and the distance measurement data that meets certain conditions is valid data. The criterion for judging the robot autonomous motion decision error from the perspective of sensor measurement data is the number of valid data on which the robot bases its autonomous motion decision within a distance of half robot length in the world coordinate system, and when this data is zero, the robot will cause a decision error.

6.5.2 *Kinematic Model of Ruin Search and Rescue Robot in Bumpy Environment*

When the robot passes obstacles of small height on a flat surface, the robot will experience a certain degree of bumps. The size of the tilt angle caused by bumps and the frequency of bumps are related to the shape and size of the obstacles and the distribution location of the obstacles. According to the shape of the obstacles, the obstacles that generate bumps can be divided into rod-like obstacles, rectangular obstacles, triangular obstacles and spherical obstacles. Now we analyze the bumpy environment with the above four kinds of obstacles respectively, and derive the function relationship between the tilt angle of the robot and the relative position of the robot and the obstacles, the size of the obstacles, and establish the mathematical model of the robot attitude in the bumpy environment.

- Pole-like obstacles

As shown in Fig. 6.13, A and B are the two endpoints of the rod-like obstacle, and P and Q are the vertices of the contact part with the ground when the robot is placed on the horizontal ground. When the midpoint of PQ is in contact with point A, the tilt angle of the robot does not exceed 40 degree, then the obstacle is regarded as an overturnable obstacle, otherwise it is an unturnable obstacle.

Take the position of point P at the initial moment of contact between the robot and the obstacle as the coordinate origin and establish the coordinate system. Let $PQ = l$, $AB = h$, the θ is tilt angle of the robot, $\theta = \angle QPB$ $OP = x'$, $OQ = z'$.

The main processes of robot overcoming obstacles can be divided into the 4 kinds of processes shown in Fig. 6.13 according to the different contact points between the robot and the obstacle.

Process 1: point P is in contact with the ground and point Q is in contact with the obstacle, the tilt angle function of the robot during the process is as follows.

Fig. 6.13 Pole-like obstacle

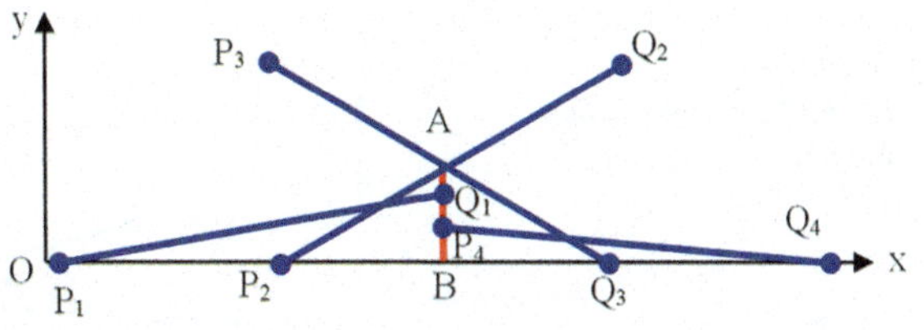

$$\theta = \arccos\frac{l - x'}{l} \tag{6.10}$$

Process 2: point P is in contact with the ground and the robottom of the robot is in contact with the obstacle, the tilt angle function of the robot in this process is as follows.

$$\theta = \arctan\frac{h}{l - x'} \tag{6.11}$$

Process 3: point Q is in contact with the ground and the robottom of the robot is in contact with the obstacle, the tilt angle function of the robot in this process is as follows.

$$\theta = \arctan\frac{h}{z' - l} \tag{6.12}$$

Process 4: point P is in contact with the obstacle and point Q is in contact with the ground, the tilt angle function of the robot during the process is as follows.

$$\theta = \arccos\frac{z' - l}{l} \tag{6.13}$$

- Rectangle-like obstacle

As shown in Fig. 6.14, A, B, D, C are four end points of a rectangle-like obstacle, and P and Q are the vertices of the contact part with the ground when the robot is placed on the horizontal ground, respectively. When the midpoint of PQ is in contact with point A, the tilt angle of the robot does not exceed 40°, then the obstacle is regarded as an overturnable obstacle, otherwise it is an unturnable obstacle.

Take the position of point P at the initial moment of contact between the robot and the obstacle as the coordinate origin and establish the coordinate system. Let PQ = l, AB = h, BD = d, the tilt angle θ of the robot, θ = ∠QPB, OP = x′, OQ = z′.

The main processes of robot overcoming obstacles can be divided into the 4 kinds of processes shown in Fig. 6.14 according to the different contact points between the robot and the obstacle.

Process 1: point P is in contact with the ground and point Q is in contact with the obstacle, and the tilt angle function of the robot during this process is as follows.

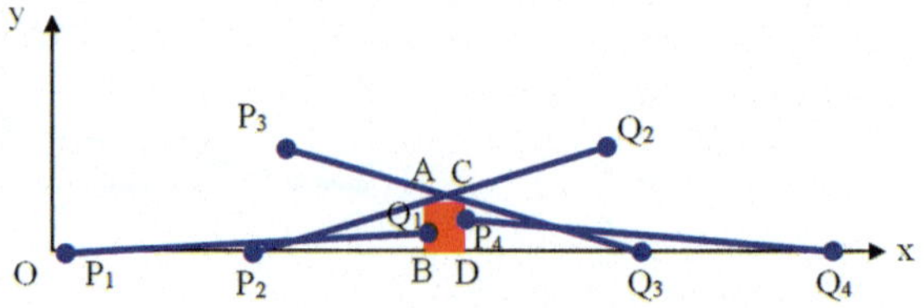

Fig. 6.14 Rectangle-like obstacle

$$\theta = \arccos\frac{l - x'}{l} \tag{6.14}$$

Process 2: point P is in contact with the ground and the robottom of the robot is in contact with the obstacle, the tilt angle function of the robot during the process is as follows.

$$\theta = \arctan\frac{h}{l - x'} \tag{6.15}$$

Process 3: point Q is in contact with the ground and the robottom of the robot is in contact with the obstacle, the tilt angle function of the robot in this process is as follows.

$$\theta = \arctan\frac{h}{z' - l - d} \tag{6.16}$$

Process 4: point P is in contact with the obstacle and point Q is in contact with the ground, the tilt angle function of the robot during the process is as follows.

$$\theta = \arccos\frac{z' - l - d}{l} \tag{6.17}$$

- Triangle-like obstacle

As shown in Fig. 6.15, A, B, C are three end points of a triangle-like obstacle, AB = AC, D is the midpoint of BC, P and Q are the vertices of the contact part of the robot with the ground when it is placed on the horizontal ground, respectively. When the midpoint of PQ is in contact with point A, the robot has only one point in contact with the obstacle, and the tilt angle of the robot does not exceed 40°, the obstacle is regarded as the obstacle that causes bumps, otherwise, the obstacle can be regarded as a slope.

Take the position of point P at the initial moment of contact between the robot and the obstacle as the coordinate origin and establish the coordinate system. Let $PQ = l$, $AD = h$, $BD = CD = d$, the tilt angle θ of the robot, $\theta = \angle QPB$, $OP = x'$, $OQ = z'$.

The main processes of robot overcoming obstacles can be divided into the 4 kinds of processes shown in Fig. 6.15 according to the different contact points between the robot and the obstacle.

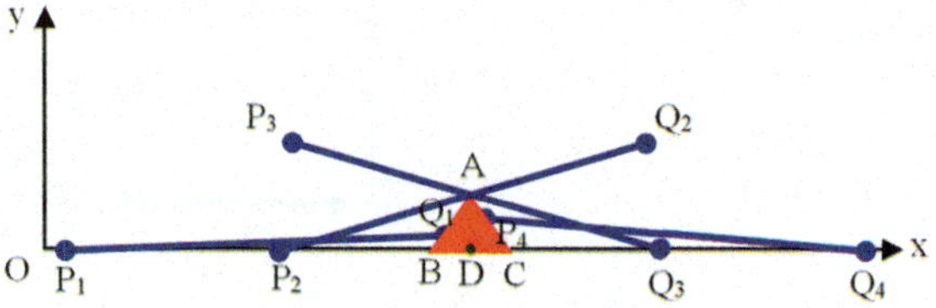

Fig. 6.15 Triangle-like obstacle

Process 1: point P is in contact with the ground and point Q is in contact with the obstacle, and the tilt angle function of the robot during this process is as follows.

$$\theta = \arccos\frac{h^2(l - x') + d\sqrt{l^2h^2 + l^2d^2 - h^2(l - x')^2}}{lh^2 + ld^2} \tag{6.18}$$

Process 2: point P is in contact with the ground and the robottom of the robot is in contact with the obstacle, the tilt angle function of the robot during the process is as follows.

$$\theta = \arctan\frac{h}{l + d - x'} \tag{6.19}$$

Process 3: point Q is in contact with the ground and the robottom of the robot is in contact with the obstacle, the tilt angle function of the robot in this process is as follows.

$$\theta = \arctan\frac{h}{z' - l - d} \tag{6.20}$$

Process 4: point P is in contact with the obstacle and point Q is in contact with the ground, the tilt angle function of the robot during the process is as follows.

$$\theta = \arctan\frac{h^2(z' - l - 2d) + d\sqrt{l^2h^2 + l^2d^2 - h^2(z' - l - 2d)^2}}{lh^2 + ld^2} \tag{6.21}$$

- Spherical obstacles

As shown in Fig. 6.16, A, B, C are three points of a sphere-like obstacle, BAC is a semicircle, BC is the diameter, D is the midpoint of BC, AD⊥BC, P and Q are the vertices of the contact part of the robot with the ground when it is placed on the horizontal ground, respectively. With a straight line connecting the three points of A, B, C to form a triangle, when the midpoint of PQ and point A contact, and P point in the horizontal coordinate axis, if the line segment PQ only has one point contact with triangle ABC, and the robot tilt angle does not exceed 40°, then the arc-shaped obstacle is regarded as the obstacle that causes bumps, otherwise, the obstacle can be regarded as a slope.

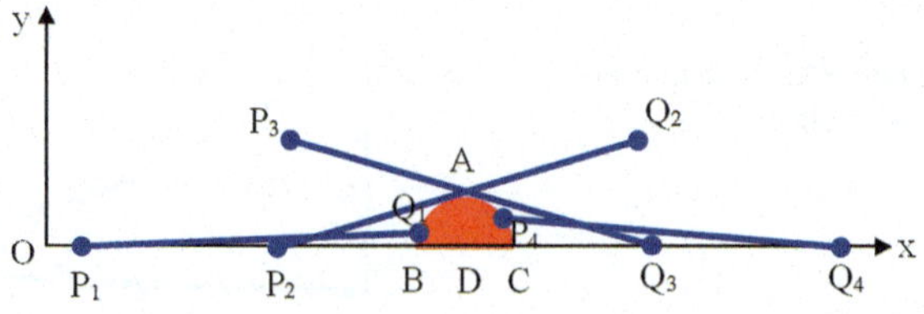

Fig. 6.16 Sphere-like obstacles

Take the position of point P at the initial moment of contact between the robot and the obstacle as the coordinate origin and establish the coordinate system. Let $PQ = l$, $AD = h$, $BD = CD = d$, the tilt angle θ of the robot, $\theta = \angle QPB$, $OP = x'$, $OQ = z'$.

The main processes of robot overcoming obstacles can be divided into the 4 kinds of processes shown in Fig. 6.16 according to the different contact points between the robot and the obstacle.

Process 1: point P is in contact with the ground and point Q is in contact with the obstacle, and the tilt angle function of the robot during this process is as follows.

$$\theta = \arccos\frac{-b + \sqrt{b^2 - 4ac}}{2a} \tag{6.22}$$

In the formula.

$$a = 4d^2l^2 - 4l^2(l - x')^2$$

$$b = 4[l^2 + (l - x')^2 - 2d^2](l^2 - lx')$$

$$c = 4d^2(l - x')^2 - [l^2 + (l - x')^2]$$

Process 2: point P is in contact with the ground and the robottom of the robot is in contact with the obstacle, the tilt angle function of the robot during the process is as follows.

$$\theta = \arcsin\frac{d}{l + d - x'} \tag{6.23}$$

Process 3: point Q is in contact with the ground and the robottom of the robot is in contact with the obstacle, the tilt angle function of the robot in this process is as follows.

$$\theta = \arcsin\frac{d}{z' - l - d} \tag{6.24}$$

Process 4: point P is in contact with the obstacle and point Q is in contact with the ground, the tilt angle function of the robot during the process is as follows.

$$\theta = \arccos\frac{-b + \sqrt{b^2 - 4ac}}{2a} \tag{6.25}$$

In the formula.

$$a = 4d^2l^2 - 4l^2(z' - l - 2d)^2$$

$$b = 4[l^2 + (z' - l - 2d)^2 - 2](lz' - l^2 - 2ld)$$
$$c = 4d^2(z' - l - 2d)^2 - [l^2 + (z' - l - 2d)^2]$$

6.5.3 Analysis and Design of Fuzzy Controller in Bumpy Environment

The task of the fuzzy controller is to control the motion speed of the robot according to the degree of bumps in the robot motion environment, to coordinate the motion speed of the robot with the degree of bumps in the environment, to improve the amount of valid data on which the robot bases its autonomous motion decision within a half-robot length distance in the world coordinate system, and then to reduce the robot's autonomous motion decision error, to prevent the robot from making a decision error, and to take into account the robot search and rescue efficiency of search and rescue work.

- Input and output variables

The input variable of the fuzzy controller is the effective sampling ratio. The system uses the effective sampling ratio of the robot range sensor measurement data as a parameter to characterize the bumpiness of the bumpy environment, and the smaller the effective sampling ratio of the measurement data, the higher the bumpiness of the bumpy environment. According to the human activity experience, the bumpiness of the bumpy environment is divided into five categories: no bumpiness, mild bumpiness, moderate bumpiness, high bumpiness and extreme bumpiness, and the bumpiness gradually increases, and the fuzzy subset of the input variables is {BN,BS,BM,BH,BE}. As shown in Fig. 6.17a, the form of the affiliation function of the input variables is shown, where the horizontal axis is the effective sampling proportion of the distance measurement data and the vertical axis is the affiliation degree.

The control objective of the fuzzy controller is to control the motion speed of the robot based on the bumpiness of the bumpy environment to reduce the possibility of

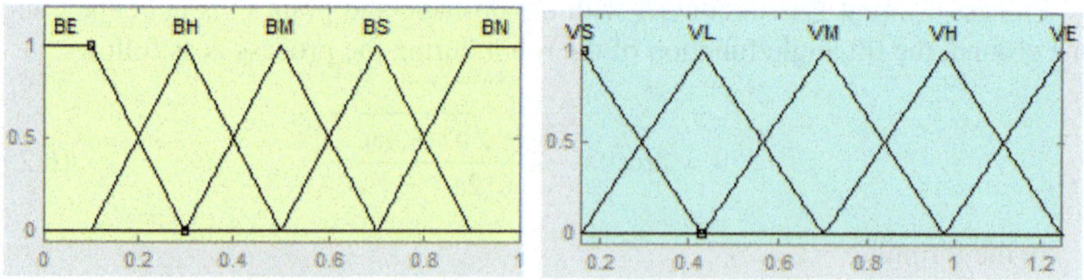

(a) Fuzzy input set (b) Fuzzy output set

Fig. 6.17 Form of affiliation function of input and output variables

Table 6.2 Linguistic description of fuzzy rules

Variables	Fuzzy subsets				
Ratio (if)	BE	BH	BM	BS	BN
V (then)	VS	VL	VM	VH	VE

robot decision errors. Therefore, the number-out variable of the fuzzy controller is the motion speed of the robot. Based on the human activity experience, the motion speed of the robot is classified into the following five categories: slow, low, medium, high and very high speed, and the output variable fuzzy subset is {VS,VL,VM,VH,VE}. As shown in Fig. 6.17b, the form of the affiliation function of the output variables is shown, where the horizontal axis is the robot motion speed and the vertical axis is the affiliation degree.

- Fuzzy control rules

According to the analysis of the input and output variables of the fuzzy controller, the input variable of the fuzzy controller is the effective sampling ratio and the output variable is the robot motion speed, therefore, the fuzzy controller is a single-input, single-output controller, and the fuzzy control rules are shown in Table 6.2, where Ratio is the input variable of the fuzzy controller, i.e., the effective sampling ratio, and V is the number-out variable of the fuzzy controller, i.e. the motion speed of the robot.

The fuzzy relation R from the effective sampling proportional domain Ratio to the velocity domain V is as follows.

$$R = (BE\hat{\times}VS)(BH\hat{\times}VL)(BM\hat{\times}VM)(BS\hat{\times}VH)(BN\hat{\times}VE) \tag{6.26}$$

where $\hat{\times}$ is the fuzzy direct product operator.

The fuzzy control matrix of the fuzzy controller is a mathematical description of the fuzzy rules. According to the affiliation functions of the input variables of the fuzzy controller and the output variables of the fuzzy controller in Fig. 6.17 and the fuzzy control rules shown in Eq. (6.26), the fuzzy control rule matrix of the fuzzy controller can be obtained as follows

$$R = \begin{bmatrix} 1 & 0.5 & 0 & 0 & 0 \\ 0.5 & 0.5 & 0.5 & 0.5 & 0 \\ 0 & 0.5 & 1 & 0.5 & 0 \\ 0 & 0.5 & 0.5 & 0.5 & 0.5 \\ 0 & 0 & 0 & 0.5 & 1 \end{bmatrix} \tag{6.27}$$

The key control role of the fuzzy controller is to control the robot's motion speed V based on an effective sampling ratio characterizing the degree of bumps in the bumpy environment, and the output variables of the robot can be obtained through the following relationship.

$$V = R \circ \text{Ratio} \tag{6.28}$$

where $\circ$ is the fuzzy set synthesis operator.

6.5.4 Simulation Research

The system was studied by simulation using MATLAB to verify the accuracy and effectiveness of the system.

- Simulation conditions

According to the analysis of the mathematical model of the robot pose in the bumpy environment, it is known that the robot tilt angle changes differently during the motion of the robot in the environment of obstacles of different sizes. Among them, the main obstacle that causes the robot to produce bumps is the rectangular-like obstacle. As shown in Fig. 6.18, the tilt angle change data of the robot in the process of passing through one obstacle in different sizes of rectangular-like obstacle environment, the horizontal axis is the horizontal distance between the robot and the starting point in the process of linear transport work in the world coordinate system, and the vertical axis is the tilt angle of the robot in the forward direction, when the robot tilts upward, the tilt angle of the robot is represented by a positive number and when the robot tilts downward, the tilt angle of the robot is expressed as a negative number. In Fig. 6.18a, the height of the rectangular obstacle is 1 cm and the width is 6 cm, and the maximum tilt angle of the robot in the process of motion is 3.583°; in Fig. 6.18b, the height of the rectangular obstacle is 3 cm and the width is 4 cm, and the maximum tilt angle of the robot in the process of motion is 10.8°. In Fig. 6.18c, the height of the rectangular-like obstacle is 6 cm, the width is 6 cm, and the maximum tilt angle of the robot during the motion is 22.01°; in Fig. 6.18d, the height of the rectangular-like obstacle is 12 cm, the width is 6 cm, and the maximum tilt angle of the robot during the motion is 29.99°.

According to Fig. 6.18 and the analysis of the mathematical model of the robot's pose under various types of bump-causing obstacles above, it can be seen that the robot's tilt angle in the bumpy environment is characterized by complexity and variability, and the robot's pose in the bumpy environment is more complex with the superposition of multiple obstacles in the debris environment.

Since the purpose of the simulation experiments is to verify the effectiveness and accuracy of the control method, the simulation experimental conditions should match the actual conditions. According to the analysis in Fig. 6.18 and the mathematical model analysis of the robot pose in the bumpy environment above, the necessary conditions for the MATLAB simulation experiments are as follows.

- The tilt angle of the robot varies with the horizontal distance of the robot from the starting point in the world coordinate system, and the tilt angle differs when the robot is tilted upwards and downwards.

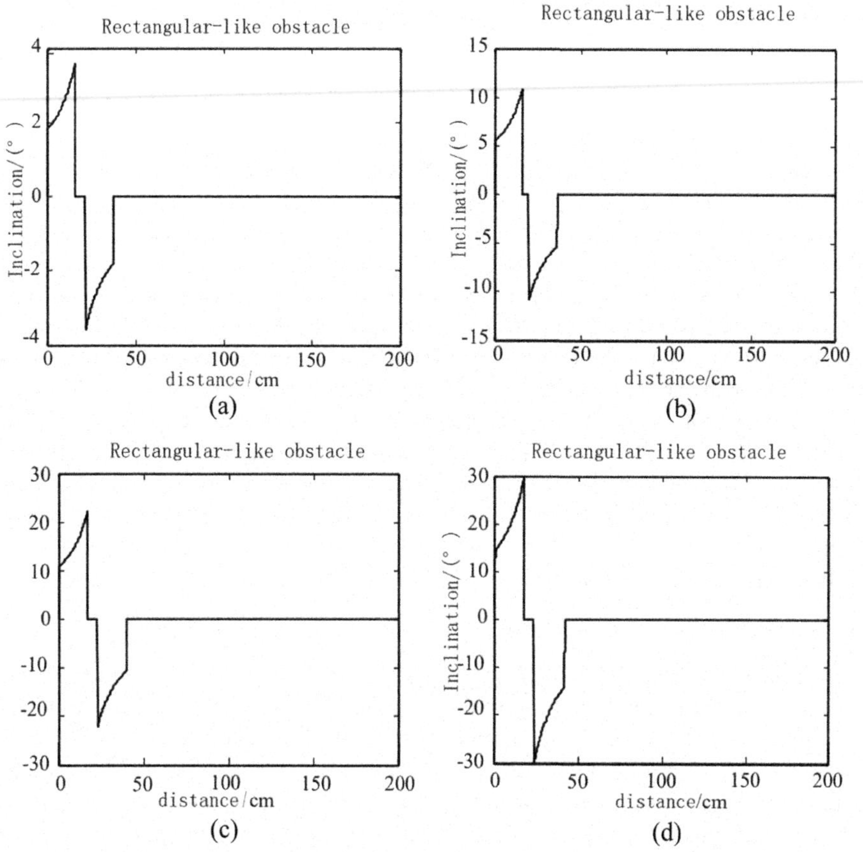

Fig. 6.18 Pose model of the robot under different obstacle situations. **a** Attitude model of the robot in the case of obstacle 1. **b** Attitude model of the robot in the case of obstacle 2. **c** Attitude model of the robot in the case of obstacle 3. **d** Attitude model of the robot in the case of obstacle 4

- The tilt angle of the robot must contain the full range of values within which the robot's tilt angle varies in the actual bumpy environment.
- The simulation conditions must include different levels of bumps and include phases of robot motion from low to high bumps, as well as phases of robot motion from high to low bumps.

Since the obstacles that cause the robot to bounce in the actual environment are mostly rectangular-like obstacles, this paper uses a simulation of the robot passing through multiple rectangular-like obstacles to obtain the tilt angle of the robot in the bouncy environment based on the analysis above. As shown in Fig. 6.19, a schematic diagram of the simulation conditions of the simulation experiment is shown. In the figure, the point O is the starting position, at this time, the robot and the obstacle began to contact, the robot on the point P and the point O overlap, the starting position for the robot and the obstacle began to contact the position, the robot in the process of

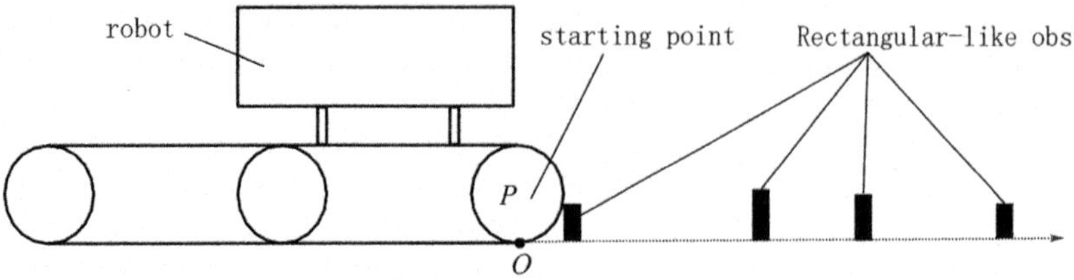

Fig. 6.19 Schematic diagram of simulation conditions

movement and the starting point between the horizontal distance that is the horizontal distance between the point P and the point O, rectangular-like obstacles should meet the above necessary conditions, with different size.

In the simulation experiment, the height of 8 rectangular obstacles with different sizes are 6, 3, 4, 12, 5, 5, 12, 2 cm, the width is 6 cm, and the distance between adjacent obstacles is 45 cm. The tilt angle is positive when the robot is tilted upward and negative when the robot is tilted downward. In this data, the tilt angle variation range includes the value range of the actual environment, and includes the data under different bumps, which satisfies the necessary conditions for the above simulation.

- Simulation results and analysis

As shown in Figs. 6.21 and 6.22, the experimental results of the simulation system are shown after the simulation data of the tilt angle of the robot under the above bumpy environment is input to the simulation system. Among them, Fig. 6.21 shows the input variables and output variables of the fuzzy controller, and Fig. 6.22 shows the number of effective sampling data obtained by the robot during the process of passing through a distance range of half of the robot length in the horizontal direction, i.e., 32 cm distance, under the control method of coordinating the motion speed with the degree of environmental bumps and and the robot using constant speed for control, respectively. The reason for choosing the distance through half of the length of the robot is that the center point of the robot is in the geometric center of the robot, and the suspension of the robot during its movement can lead to violent collisions between the robot and other objects, and even to robot tipping.

In Fig. 6.21, the horizontal axis is the horizontal distance of the robot from the starting position, and the vertical axis is the effective sampling ratio of the input variable and the motion speed of the robot of the output variable, respectively. From Fig. 6.21 and combined with the analysis of the simulation conditions in Fig. 6.20, it can be seen that in the region where the environmental bumps are large, for example, between the horizontal axis coordinates [10, 40], the tilt angle of the robot is larger, resulting in a smaller effective sampling ratio, and in this condition, the robot decreases the motion speed; conversely, in the region where the environmental bumps are smaller, for example, between the horizontal axis coordinates [320, 350], the tilt angle of the robot is smaller, the effective sampling ratio is larger, and the robot increases the motion speed. The simulation results shown in Fig. 6.21 are consistent with the theoretical analysis above.

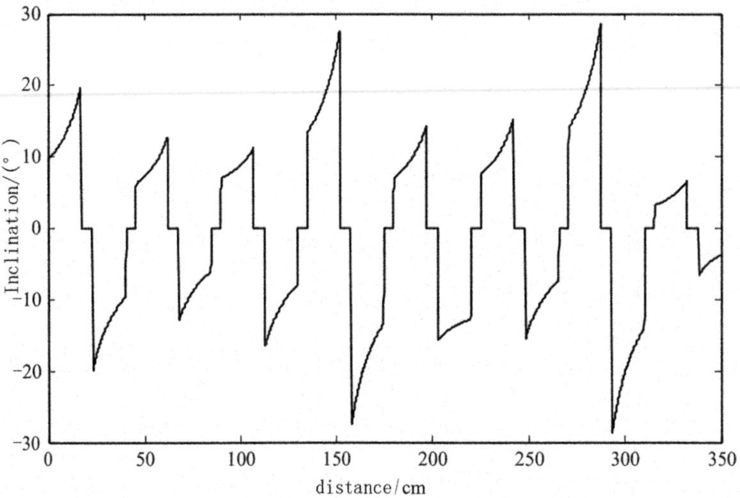

Fig. 6.20 Robot tilt angle data

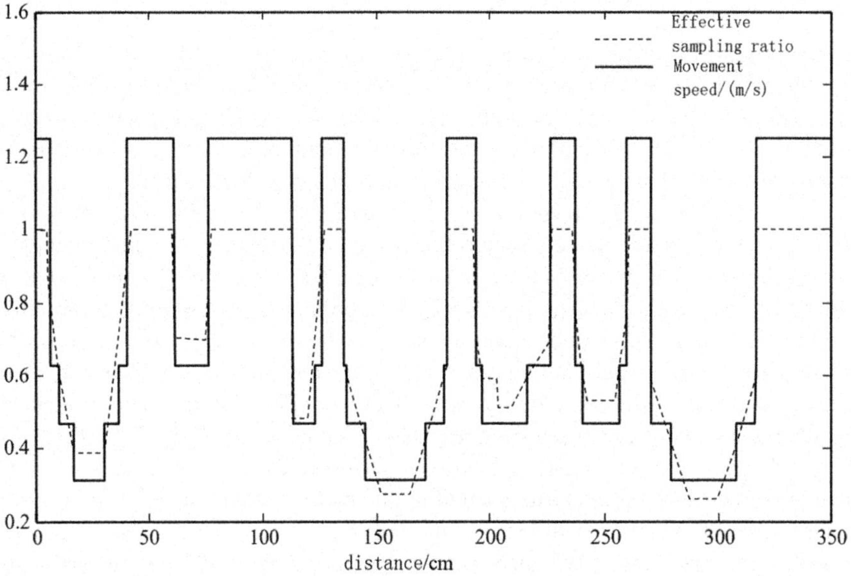

Fig. 6.21 Input variables and output variables

Figure 6.22 shows the number of effective distance sampling data of the robot during the robot passing through the same distance under the velocity coordinated control motion condition and the robot using constant velocity motion condition, respectively, and the number of effective samples in the velocity coordinated condition in the Fig. are not less than 1. As the point (170.5,3) in the figure, for the robot

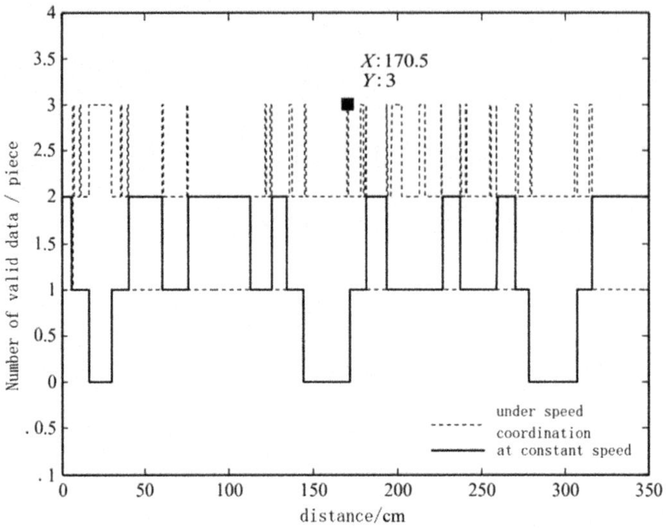

Fig. 6.22 Comparison of the number of valid data

in the process of passing the position interval of distance [138.5, 170.5] from the starting point, the number of effective distance sampling data obtained by the robot is 3. The comparison of the data in the two cases in the figure shows that in the case of bumps, when the robot adopts uniform speed motion, in the process of passing half of the robot's length, there is a situation in which the effective distance sampling data is 0, in this case, the deviation of the distance data obtained by the robot is too large, and the decision made by the robot based on the distance sampling data is a wrong decision, which causes the robot's decision error. In contrast, using a control method in which the robot motion speed is coordinated with the degree of environmental bumps, the number of valid distance sampling data in the process of the robot passing half of the robot length is 1 at least, and in this process, the robot can discard all invalid distance sampling data and rely on valid sampling data for decision making. For example, between the horizontal axis coordinates [270, 300], Fig. 6.20 shows that the robot has a large inclination angle in this interval, and Fig. 6.21 shows that the effective sampling ratio is small in this interval, which together indicates a high degree of environmental bumps, and Fig. 6.21 shows that the robot's motion speed in this interval is coordinated with the degree of bumps, and the speed is small, and Fig. 6.22 shows that when 1.02 m/s uniform speed is used, the effective sampling number is 0, there is no effective decision basis data, while using the speed coordination method, the effective sampling number is greater than 1, there is effective decision basis data. The above analysis shows that the control method of coordinating the motion speed with the degree of environmental bumps can effectively solve the problem of the robot's decision error in the bumpy environment, which verifies the effectiveness and accuracy of the control method.

At the same time, it appears in the figure that the number of effective distance sampling data in the condition of motion control using velocity coordination is lower than the number of effective distance sampling data in the case of uniform motion, resulting from the fact that the motion speed of the robot in this case is greater than the motion speed in the condition of uniform motion. The above analysis can be verified by combining Figs. 6.21 and 6.22. The above analysis shows that a control method that coordinates the motion speed with the degree of environmental bumps can effectively improve the efficiency of the robot search and rescue work.

Chapter 7
Text Questions and Answers Robot

7.1 Overview of Text Questions and Answers Robot

7.1.1 Concepts and Features of Text Questions and Answers Robot

The text questions and answers robot is an intelligent human–computer interaction system in which users ask questions in natural language and the robot finds the exact answer from a large amount of information. Text questions and answers robots are designed to allow users to ask questions in natural language and get answers directly. For example, a user asks "Where is the Chinese Academy of Sciences?" The text questions and answers robot answers "The Chinese Academy of Sciences is located at Sanlihe Road52, Xicheng District, Beijing". The Text questions and answers robot described in this paper is a Chinese Text questions and answers robot.

Text questions and answers robots belong to quiz systems, which are one of the most interested research areas at the International Conference on Text Retrieval, the highest level conference in the field of text processing.

Traditional search engines are based on keyword searches, returning a large amount of relevant information, and users find relevant information from the information returned by the system. The Text questions and answers robot directly returns the only accurate answer for the user, the answer is more concise, and the time cost for the user to obtain accurate information is lower this.

The two main performance characteristics of the Text questions and answers robot are.

- The question and answer portal is a natural language form of interrogative sentence.
- The result of the question and answer is a sentence or a paragraph directly related to the question.

T. Guo et al., *Special Robot Technology*, Advanced and Intelligent Manufacturing in China, https://doi.org/10.1007/978-981-99-0589-8_7

Thus, compared to traditional search engines, text-based question and answer robots have obvious advantages.

- Natural language questioning in a manner more consistent with human interaction habits.
- Compared to keywords, statements contain more complete information and can express user intent more accurately.
- Accurate and concise answers that directly address the user's questions, with more efficient information retrieval.
- The simple answer format makes the text-based Q&A robot more suitable for mobile Internet applications as well as IoT human–computer interaction device applications.

7.1.2 Development History of Text Questions and Answers Robots

Although as early as 60s the era of artificial intelligence just began to research, people proposed to let the computer use natural language to answer people's questions, the academic and industrial sectors began to build the prototype of the question and answer system. And until 80s the era of question-and-answer systems have been limited to special areas of expert systems, although the Turing experiment told people that if computers can talk to people like humans, it can be determined that computers have intelligence, but because of the limited conditions at the time, the experiments were conducted in restricted areas, or even on fixed paragraphs.

In recent years, with the rapid development of network and information technology, especially the popularity of mobile Internet and the development of Internet of Things (IoT) technology, mobile Internet devices have gradually become popular, and the emergence of various types of human–computer interaction devices in the IoT era has mushroomed. The traditional keyword-based search engine is difficult to meet the information retrieval needs of users in mobile Internet devices and IoT human–computer interaction devices, and the desire of people want to obtain high quality information faster has promoted the research and development of text- based Q&A robotics. Many large research institutes and well-known enterprises are actively involved in this field of research.

Among the foreign question-and-answer robot systems, the main ones are the START system developed by MIT, the AnswerBus system developed by the University of Michigan, the AskMSR system by Microsoft, and the NTCIR system by Japan.

Compared to the research progress of foreign question and answer robot systems, the research of domestic text question and answer robot systems started late, and only after the 1970 did they start to research Chinese text question and answer robot systems based on Chinese, and the Institute of Languages of the Chinese Academy of Sciences researched China's first human–computer dialogue system based on

Chinese in the 1980, and as more and more research institutions researched text question systems in China, the achievements made became more and more fruitful, Tsinghua University, Fudan University and Beijing Language and Culture University have made greater achievements in the field of Chinese natural language research. Tsinghua University has researched EasyNay, a campus navigation system, and the Q&A system developed by the Institute of Computing of the Chinese Academy of Sciences provides answers to the relationships of characters in Dream of the Red Chamber.

7.1.3 Classification of Text Questions and Answers Robots

Text questions and answers robots exist in a variety of classification methods due to the different application areas, the unique form of information storage, and the different data sources from which the robot answers come.

Common approaches to classify text-based Q&A robots include by domain, by Q&A format, and by Q&A corpus.

- Classification by domain

The division by domain refers to the division based on the domain of the content answered by the text Q&A robot, which is divided into restricted domain text Q&A robots, FAQ text Q&A robots, and open domain text Q&A robots.

– Restricted domain text Q&A robot

Restricted domain text Q&A robots refer to domain-specific text Q&A robot systems, such as medical, finance, law, education, real estate domain, whose answers are restricted to a specific domain, and the robot's corpus is constructed based on domain-specific information rather than on the Internet as a search data source. Thus, restricted domain Q&A robots have a clear and relatively fixed data source.

Restricted domain text questions and answers robots have three characteristics.

a. Fixed application domain: The corpus information source of the restricted domain Text questions and answers robot is not the data source of Internet search, but the database and knowledge base of a specific domain, and the system is pre-designed for the questions that users may ask in advance. Since it is pre-designed user questions and restricted to fixed domain corpus information, the data source of the restricted domain Text questions and answers robot must be explicit and also authoritative.
b. The system has a certain degree of complexity: restricted domain Q&A robot systems must meet all of the user's questions and answers in a particular domain. Due to the complexity of the user's questions, which leads to the complexity of restricted domain Q&A robot systems, it is difficult for the system to use a single simple algorithm and model to meet the diverse user Q&A needs.

c. Good usability: restricted domain Text questions and answers robots are for specific application areas, users have clear needs, and the system should meet the specific needs of different users in the field, thus, restricted domain Text questions and answers robots must have good usability.

Therefore, the core of the system of restricted-domain text Q&A robots is to build domain-specific Q&A corpus, and at the same time, the acquisition and representation methods of the Q&A corpus shall be different according to the industry, to build industry domain-based knowledge base with industry-specific specific knowledge and special to go, and thus the relative size of the Q&A corpus is small.

– FAQ text Q&A robot

FAQ text-based Q&A robot is an intelligent Q&A system based on a FAQ dataset, where the questions and answers data are organized in relation to each other, in a paired list composition.

The FAQ question and answer datasets are known information, and since they are built on known information, the system has no error messages and performs efficiently. the FAQ text-based Q&A robot retrieves the question set by taking the questions asked by the user, and if the target question is retrieved, the robot returns to the user the answer information that maps one-to-one to the target question.

Since FAQ Text questions and answers robots can only answer questions predetermined by the designer, the disadvantage of FAQ Text questions and answers robots is also obvious, i.e., the size of the FAQ dataset is small, and FAQ Text questions and answers robots are generally applied to a certain aspect of a restricted field, such as frequently asked questions about comprehensive administrative service halls in the government industry, financial reimbursement questions in the internal OA system of enterprises, banking APP software usage, hospital registration process and the process of booking specialists.

– Open domain text questions and answers robot

Compared with restricted-domain text-based robots, open-domain text-based robots are multi-domain oriented question and answer systems, which are significantly different from restricted-domain text-based robots in terms of data size and domain context.

In terms of data size, the corpus size of restricted domain text Q&A robots is limited to a specific domain, and the scope of its Q&A is also limited to the domain restricted by the corpus; the corpus used by open domain text Q&A robots is oriented to multiple domains, therefore, it is more technically difficult to meet the system requirements by using conventional keyword retrieval techniques, and for processing large-scale text data, natural language processing techniques must be used for data processing.

The core goal of the open-domain text-based question and answer robot is to provide introductory cross-domain question-and-answer interactions, where users ask questions through natural language and the system obtains accurate answers from various data sources. Since the corpus of the open domain Text questions and answers robot is not restricted to domains, the user's questions are not restricted to specific domains.

A typical application of open-domain Text questions and answers robot system is an open-domain Text questions and answers robot system based on Internet information, where the user talks to the robot through a simple human–computer interaction interface, and the Text questions and answers robot provides brief and accurate answer information through a large amount of Internet information.

- Classification by question and answer format

Text-based Q&A robots also differ in their Q&A format. By Q&A format, text-based Q&A robots are classified as chat robots, retrieval-based Q&A robots, and community-based Q&A robots.

– Chatting robot

Chatting robot are human–computer dialogue systems that simulate human conversations, with robots answering various questions that people ask.

The basic principle of chatting robot is to design a pattern matching approach based on conversational skills. The pattern matching model for chatting robot is relatively simple, taking the natural language interrogative sentences entered by the user and processing them in terms of words as the basic unit to find the corresponding answer to the user's question.

Since chatting robot build their algorithmic models in terms of words, the system is relatively simple in processing user questions with insufficient analysis of semantics, insufficient understanding of interrogative sentences, and weak in contextual processing. Therefore, for complex user questions or in the case of a larger question–answer corpus, the answer is prone to the problem of not answering the question.

Because of the above characteristics, chatting robot are suitable for human–computer Q&A scenarios that deal with simpler, smaller-scale problems, such as a specific group of users in an exact segment, or simple high-frequency questions in a specific part of the system.

– Retrieval-based Q&A robot

A retrieval-based Q&A robot is a kind of Q&A robot that combines a search engine with natural language processing. Users enter questions in the form of natural language, and based on their natural language interrogations, the system searches from the Internet or its document repository and feeds back to the user the search results of documents, web pages, etc. related to the user's question.

Retrieval-based Q&A robots differ from simple search engines in that the system takes the natural language interrogative sentences provided by the user and analyzes the intent of the questions asked by the user through processing such as interrogative sentence analysis and question understanding to retrieve the data source. Traditional search engines, on the other hand, are mainly based on keywords for retrieval and lack sufficient semantic understanding of the intent of the user's questions and interrogative sentences.

Retrieval-based Q&A robots also have shortcomings in information retrieval capabilities compared to search engines. After a long period of research and system iteration and upgrade, the international and domestic excellent search engine systems have been equipped with quite powerful functions, and the accuracy and recall rate of search results are more satisfactory to users when searching for the massive amount of data on the Internet, and their information retrieval ability has exceeded the ability of question-and-answer type Q&A robots.

– Community Q&A robot

Community Q&A robots, also known as collaborative Q&A systems, are an Internet-based, open-domain Q&A system. The Q&A corpus of a community Q&A robot is sourced from Internet users, who ask questions in a natural language way, and the community Q&A robot retrieves the best answer in the corpus through information retrieval and feeds it back to the user.

The community Q&A robot system has obvious social network attributes, and its biggest feature is that it attracts many Internet users to participate in the process of asking questions and proposing answers, and builds a gradually improving corpus through the interaction and collaboration of different user groups. Due to this feature, the corpus contributors of community Q&A robots bring together the wisdom of various industries within the community Q&A robot system, gradually developing large systems such as Baidu Know, Sina AiQ, and Zhihu.

The rich question and answer corpus of the community Q&A robot system constitutes a large-scale dataset, which provides new resources and ways to study natural language processing, information retrieval, information extraction, machine learning, and big data. It is a challenging and expected topic to mine more valuable and meaningful information from these datasets.

- Classification by question and answer corpus

Question and answer corpus is an important and indispensable part of a text-based Q&A robot system. The question and answer corpus can be divided into structured data (e.g., relational database), semi-structured data, and unstructured data (e.g., web pages). According to the different question and answer corpus of Text questions and answers robots, Text questions and answers robots can be divided into structured database based Text questions and answers robots, free text based Text questions and answers robots and knowledge base based Text questions and answers robots.

– Text-based Q&A robot with structured database

Structured databases, which also become row databases, are data logically expressed and realized by a two-dimensional table structure, strictly following data format and length specifications, and are stored and managed mainly through relational databases.

The main feature of the text-based Q&A robot system based on a structured database is that the system takes the user's question as a query condition, analyzes the user's question, performs the query operation in the structured database, and gives the query result back to the user as the answer.

Traditional structured database query requires query statements to strictly follow query conditions and specific formats, and if a user is not able to understand structured databases very well, it is difficult for traditional database systems to execute that user's query operations and even more difficult to get accurate query results. Therefore, the key to a structured database based text-based Q&A robot system is to take the questions described by the user's natural language, understand and analyze them, and convert the natural language into the form of structured database query language accurately and efficiently, and then query the structured database data.

– Free-text based text Q&A robots

Free text is raw, unprocessed, unstructured text, and documents, web pages, etc. are all free text.

The free-text based question and answer robot system allows users to ask questions in natural language, and the system retrieves documents and web page data matching the user's question from the system's free-text collection or the Internet through information retrieval, and then extracts the answer to the question from the retrieved text or web page and feeds it back to the user through answer extraction.

Answers to questions that can be answered by free-text-based text-based Q&A robots exist in documents, web pages, and other systems. Since these free-texts have no domain restrictions, free-text based Q&A robots are mostly open-domain Q&A robots, which include systems such as community Q&A robots for Internet applications.

– Knowledge-base based text Q&A robots

A knowledge base is a tool for processing information, a system for producing, processing and storing complex structured and unstructured information. The first generation knowledge base system is an expert system. The process of information processing and processing is the process of knowledge base creation and application, and the processing and application of knowledge base includes information processing, handling, storage, retrieval and application. The two pillars of the knowledge base include AGENT and Ontology.

A knowledge-base based text-based Q&A robot uses a knowledge base to answer questions posed by users, and the knowledge base is an important component on which the robot relies for support. A knowledge-base based text Q&A robot can use one or more knowledge bases to understand and solve the user's Q&A questions using techniques such as retrieval and inference.

Knowledge-base based Text questions and answers robots have a high accuracy rate due to the use of a knowledge base that has been processed with information to refine and sublimate the raw data.

7.1.4 Evaluation Metrics for Text Questions and Answers Robots

The performance evaluation of text-based question and answer robots is generally evaluated using two metrics: accuracy (precision) and recall, and the correctness and recall represent the overall performance of the whole system.

Accuracy is the ratio of the number of correct messages extracted to the total number of messages extracted; recall is the ratio of the number of correct messages extracted to the total number of messages in the sample.

For example, the number of knowledge in the answer base of a Text questions and answers robot is A. The robot matches m knowledge based on the user's question, of which m1 questions are correct questions and m2 questions are incorrect questions, while there are actually n knowledge in the answer base that match the user's question, at which point the accuracy is m1/n and the recall is m1/A.

7.2 Architecture of Text Questions and Answers Robot

7.2.1 Basic Principles

The functional representation of a text question and answer robot is the process of finding the corresponding answer based on the user's question. The mathematical description of a text question and answer robot is the process of solving some element a of the answer set A for a known system of question set Q, answer set A, mapping relation F, and some element q of the question set.

For example, if a user asks "Where is the Chinese Academy of Sciences?" In order to understand the user's intention, the robot first analyzes the user's question, and the question analysis shows that the user is asking about a location, and the name of this location is "Chinese Academy of Sciences" (full name "Chinese Academy of Sciences"). Then, the system extracts the answers in the answer database, and only extracts the geographic location in the answer database as the candidate answer, and finally selects the top information "Chinese Academy of Sciences is located at Sanlihe Road52, Xicheng District, Beijing" from many candidate answers as the answer to the user.

Thus, the Text questions and answers robot system consists of three main components: question analysis, information retrieval and answer extraction (Fig. 7.1).

Question analysis is the decomposition and analysis of user questions, generally including lexical analysis, syntactic analysis, question type judgment, syntax judgment, named entity identification and other processes, and the results of question analysis prepare data for information retrieval, and also serve for answer extraction.

Information retrieval is similar to information retrieval in search engines, where the purpose of information retrieval is to search a data set such as a database, knowledge base or web page for all information that may contain an answer based on the

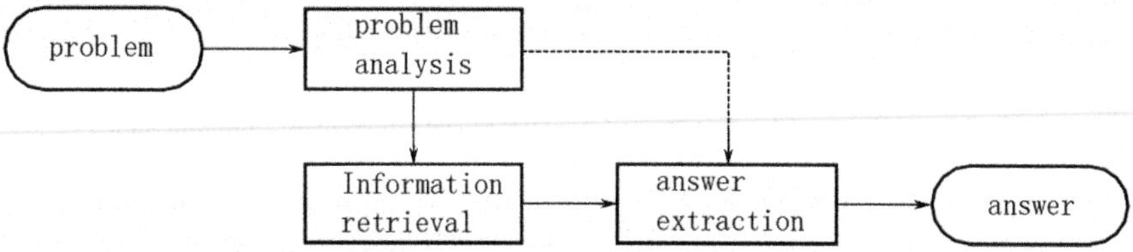

Fig. 7.1 Basic principle of Text questions and answers robot

query criteria, and to recall the matched target information by initial filtering based on the criteria. The results of information retrieval are processed by the answer extraction stage for further analysis.

Answer extraction is one of the core aspects of text-based question and answer robots. The main goal of answer extraction is to extract information that matches the user's question from the information provided by information retrieval and feed it back to the user as the final answer of the robot. The key to answer extraction is to analyze the results of information retrieval and match them with the analysis results in the question analysis stage to obtain the answers contained in the information retrieval results.

The purpose and role of each part of question analysis, information retrieval, and answer extraction in the text question and answer robot system, the processing objects, key technologies, and output results are shown in Table 7.1.

At the same time, in order to improve the information retrieval efficiency and accuracy of the Text questions and answers robot, the Text questions and answers robot also contains a frequently asked questions (FAQ) library, which extracts, analyzes and stores the high-frequency questions frequently asked by users and the corresponding answers. After the system obtains the user's question, it first searches through the FAQ library, and if the FAQ contains a question matching the user's question, the system directly gives the corresponding answer without going through the information retrieval and answer extraction process (Fig. 7.2).

7.2.2 System Architecture

The overall system architecture of Text questions and answers robots will be partially different depending on their categories, but the basic principles of processing information are the same for all types of Text questions and answers robots. The system architecture of a Text questions and answers robot system affects not only the performance metrics such as accuracy and recall of the system, but also the non-functional metrics such as security, usability, and scalability of the system.

To ensure the integrity, high availability, and high reliability of the Chinese Q&A robot system, the system architecture must meet the following conditions.

- Integrity

Table 7.1 Relationship between question processing, information retrieval, and answer extraction

Project	Problem handling	Information retrieval	Answer extraction
Purpose and role	Question parsing, serving for subsequent processing	Get documents or web pages that may contain answers, and provide processing objects for answer extraction	Judging and generating answers from the results obtained from information retrieval
Processing object	Questions asked by users	Parsed data from problem processing	Retrieved and preselected documents or sentences
Output result	Question sentence formalization and the keyword sequence after formalization expansion	Retrieved and preselected documents or sentences	Generate answers to object questions
Key technology	Lexical analysis, syntactic analysis, problem classification, named entity recognition, sentence pattern recognition, semantic analysis, corpus technology, etc.	Boolean retrieval technology, vector retrieval model, concept retrieval model, search engine technology, etc.	Named entity recognition, syntactic analysis, similarity calculation, semantic analysis, pattern matching, statement generation, etc.
Impact on the overall system	The basic and core part of the system, the processing results affect the performance of the overall system	The processing result affects the corresponding speed of the system, and the recall rate affects the quantity and quality of the data in the answer extraction stage, which in turn affects the accuracy of the overall system	The core and goal of the system depends on the results of problem processing and information retrieval

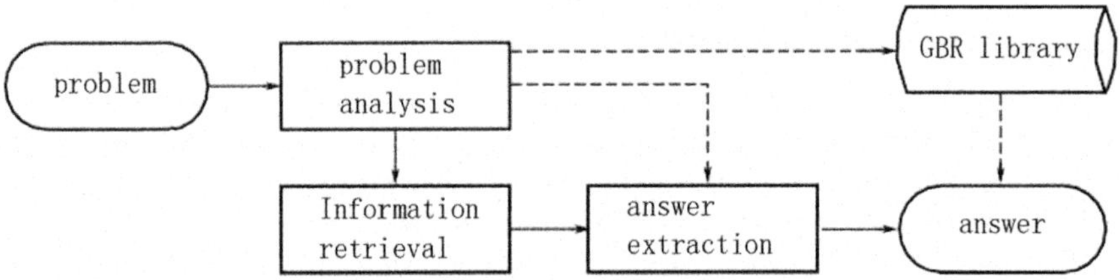

Fig. 7.2 Fundamentals of a text-based Q&A robot with a FAQ library

The system is able to perform a complete analysis of all aspects of the Chinese Q&A robot system, covering the complete aspects of the system such as question analysis, information retrieval, and answer extraction.

- Versatility

The overall system architecture can be applied to different fields, different question and answer forms, and different data sources of Chinese question and answer robot system.

- High availability

The system has a strong information processing capability, making full use of natural language processing technology to deeply analyze the unique properties of data such as questions and answers.

- Security

The system conforms to the standard system specifications and provides corresponding data security and system security for different stages and levels.

- High reliability

The system must meet the needs of long-term uninterrupted operation and be able to recover quickly in the event of a system error, and at the same time, reduce the impact of a partial system error on the overall system.

- Scalability

The system can meet the needs of technology upgrade and system performance upgrade, and carry out technology upgrade expansion and performance upgrade expansion, and the system processing process is weakly coupled with key technologies, and technology upgrade and performance upgrade do not need to change the system architecture (Fig. 7.3).

7.2.3 Problem Analysis

Question analysis is one of the foundations and core of the text-based Q&A robot system, and is the initialization module of the system, which provides in-depth analysis and understanding of the information asked by users. The input information of question analysis is the original data of user's question, and the question analysis part needs to complete several parts of work such as question type analysis, analysis of syntactic structure relationship of question, question keyword extraction and keyword expansion.

- Lexical analysis

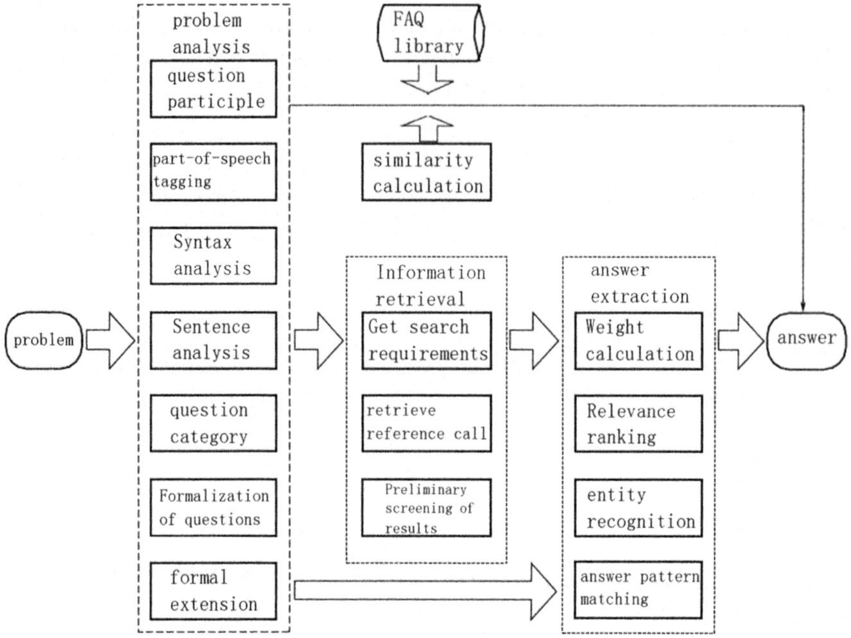

Fig. 7.3 Text questions and answers robot Architecture

Lexical analysis is the process of converting user questions into word sequences. Lexical analysis starts with word division, identifying different words according to word formation rules, and then lexical annotation, marking the lexical nature of each word in the division result to determine whether each word is a noun, verb, adjective or other lexical nature.

- Problem classification

The common question classification in Chinese contains question types such as time, place, person, reason, number, etc. For different question types, the Text questions and answers robot can develop corresponding answer extraction rules to ensure that the system obtains answers to questions according to the answer extraction rules in the answer extraction stage. At the same time, question classification can also be based on simple questions, factual questions, definitional questions, summary questions, inferential questions, etc.

- Keyword extraction

Keyword extraction is the process of extracting effective keywords from the question. In the process of keyword extraction, keywords are extracted and filtered for different word types, and deactivated words such as "ah", "bar" and "what", which have less influence on the intention of the question, are filtered out. The keywords were extracted from nouns, verbs, and adjectives, which have a greater influence on the intention of the question.

- Keyword expansion

Since there are synonyms and polysemous words in Chinese, for example, "China", "our country" and "the whole country" are synonyms, the keywords in the question and answer statements are synonyms. Therefore, the keywords in the question statement and the answer statement are synonyms, which leads to the problem of losing the data containing the correct answer due to the failure of keyword matching. Therefore, keyword expansion is required. Keyword expansion can improve the recall rate of the Q&A robot system, but there is a risk of reducing the accuracy rate.

- Syntactic structure analysis

Syntactic structure analysis is to analyze the dependency and logical structure of the words in the question and extract the main components of the user's question through syntactic structure analysis. The syntactic structure analysis lays the foundation for information retrieval and answer extraction.

7.2.4 Information Search

Information retrieval is to use the keyword sequences in the question analysis results to find the information that meets the retrieval criteria in the document collection or Internet web pages, and if the system has a FAQ library, the system also has to search in the FAQ library again. Information retrieval is the intermediate link of the system, connecting the question analysis and answer extraction links, with a bridging role.

The input information of information retrieval is the keyword sequence, i.e., the result of the problem analysis. The output information of information retrieval is the answer sets such as document sets, paragraph sets or statement sets that satisfy the retrieval conditions. The key to information retrieval is to calculate the correlation between retrieval conditions and retrieval results, determine the weights of answer set elements according to the correlation, and rank the answer set elements, obtain the answer set elements with the greatest weights, and transmit them to the answer extraction link for further processing.

Information retrieval requires indexing of the retrieved information to ensure that the system can quickly find the answer set elements containing specific keywords. At the same time, pre-processing such as invalidation deletion and de-duplication of information is required before indexing is constructed.

Information retrieval technology is a more traditional technology, and currently has more mature information retrieval models, such as Boolean retrieval, vector retrieval, concept retrieval and other retrieval models.

7.2.5 Answer Extraction

Answer extraction is the last step of a text-based Q&A robot, which is the process of refining the results of information retrieval into a final answer, and feeding the final answer back to the user. Answer extraction synthesizes the output results of question analysis and information retrieval, extracts useful information from the information retrieval output, and makes a conclusive output of the answer to the question.

In the answer extraction process, firstly, according to the question classification results in the question analysis results, the irrelevant answers are filtered out based on the filtering mechanism; then, according to the retrieval results in the information retrieval stage, the location information of the answer information is obtained through the operations of paragraph breaking, removing question sentences, filtering answer sentences, naming entity identification and sorting, etc., and a collection of answers containing multiple candidate answers is obtained. Finally, the weight of the candidate answers in the answer set is calculated, and the candidate answer with the largest weight is the final answer fed to the user.

7.3 Key Technologies of Text Questions and Answers Robots

The key technologies used in Text questions and answers robots differ in the question analysis, information retrieval, and answer extraction stages, with different objectives at different stages. The key technologies of Text questions and answers robots contain Chinese word separation technology, lexical annotation technology, deactivation word technology, feature extraction technology, question classification technology, answer extraction technology, etc.

7.3.1 Chinese Word Separation Technology

Words are the basic units that constitute the intention of an utterance. One of the biggest differences between Chinese and English is that English words are separated from each other by spaces, while Chinese text is continuous between words, therefore, the first step of Chinese natural language processing is Chinese word separation. The main function of Chinese word separation is to label all words in an utterance with their corresponding labels.

The three basic issues of Chinese word separation are word specification, ambiguous cut-off and identification of unregistered words.

- Sub-word specification

For example, the term "research biology" includes the words "research", "graduate student", "creature" and "biology". and there are many different results depending on the way the words are defined.

- Ambiguity cut

There are often ambiguous words in Chinese, for example, "research biology" can be "graduate student /thing/science" or "research/biology/science", The ambiguity cut is to cut the ambiguous words to make a judgment. The ambiguity cut is usually combined with the context, even the tone, pause, etc.

- Unregistered word recognition

Unregistered words are words that are not included in the word list or have not appeared in the training process. For newly emerged common words, new word discovery technology is used to mine and discover unlisted words, and add them to the word list after verification; for proper nouns outside the word list, nomenclature identification technology is used to separately identify names of people, places and units.

Chinese word separation technology has been studied for a long time and has more mature word separation algorithms, at present, the accuracy of better word separation systems has exceeded 90%. There are four commonly used Chinese word separation methods: word list-based word separation method, semantic analysis-based word separation method, statistical model learning-based word separation method and deep learning neural network-based word separation method.

- Word list-based word separation methods

The word list-based approach is to match the phrase to be analyzed with the words in the word list according to a certain strategy, and the match is successful if the matching word is found in the word list. It is the first Chinese word separation method that relies on word lists.

Currently, there are word list-based word separation methods such as forward maximal matching method, reverse maximal matching method, two-way scanning method, word-by-word traversal method, and n-gram word separation method.

– Forward maximum matching method

The forward maximum matching method is related to the number of characters contained in the longest word in the word table. The system takes m characters of the statement to be analyzed from left to right as the matching field, m is the number of characters contained in the longest word in the word table, and then matches the matching field with the word table. If the match is unsuccessful, the last word of the matching field is removed and the remaining characters are used as the new matching field, and the above matching process is repeated until all words are cut out. For example, the longest word in the word list contains 4 Chinese characters, in the process of word separation for "research biology", first match "research biology" with the word list, if the match is unsuccessful, then the word “graduate student” is matched with the word list, and if the match is successful, the word “material science” is matched with the word list, until all the words are cut.

– Reverse Maximum Matching Method

The reverse max-match method is the inverse of the forward max-match method. If the match is unsuccessful, the system removes the previous word from the matching field and proceeds to the next round of matching.

– Two-way scanning method

The two-way scanning method, also known as the two-way maximum matching method, compares the results obtained by the forward maximum matching method and the reverse maximum matching method, and then determines the appropriate word separation method.

– Word-by-word traversal method

Word-by-word traversal method, also known as word-by-word matching method, is a word-by-word matching method based on an index tree, where each word to be matched is matched in turn and synchronously from the root node of the index tree. This method has the advantage of fast execution efficiency, but the disadvantage is that it is complicated to build and maintain the index tree.

– n-gram syllogism

The n-gram word separation method is a Bayesian statistics-based word separation method. The system first performs word separation according to different word separation methods, at which time the separation results contain the ambiguity cut and unregistered word problems, and then constructs a directed acyclic graph with words as nodes and conditional probabilities as edges to transform the word separation problem into a problem of solving the best path, as shown in Fig. 7.4. n-gram word separation method counts the frequency of occurrence of each word in the word table, and all possible subword results are counted and the subword result with the highest probability is calculated. The joint probability of each word is shown in Eq. (7.1). n-gram word division is the same for each word a conditional probability that depends on all the words before it, and n is greater 4 than that, which will lead to data sparsity problem. 2-g word division is generally the common word division method. 2-g word division joint probability is shown in Eq. (7.2).

$$p(\omega_1, \omega_2 \ldots \omega_n) = p(\omega_1)p(\omega_2|\omega_1)p(\omega_3|\omega_1\omega_2)\ldots p(\omega_n|\omega_1\omega_2\omega_{n-1}) \quad (7.1)$$

$$\mathrm{p}(\omega_1, \omega_2 \ldots \omega_n) = \prod_{i=1}^{n} \mathrm{p}(\omega_i|\omega_{i-1}) \quad (7.2)$$

where ω_i is the i-th word in an utterance of length N.

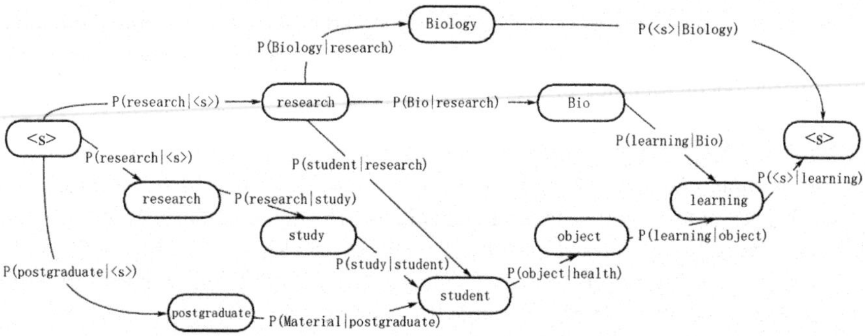

Fig. 7.4 n-gram word splitting method

Due to the strong reliance on word lists, word list-based word separation methods are less effective in semantic ambiguity and unregistered word processing.

- Semantic analysis-based word separation method

Semantic analysis based word separation methods introduce semantic analysis to process more linguistic information of natural language for word separation. The common ones are expanded transfer network method, matrix constraint method, etc.

– Expanded Transfer Network Method

The approach is based on the concept of a new state machine. The first expansion of the finite state machine, which can only recognize regular languages, makes it recursive, forming a recursive transfer network (RTN). In the RTN, the flags on the arcs can be not only final (words in the language) or non-final (word classes), but can also invoke additional sub-network names for non-final characters (e.g., word or string formation conditions). In this way, the computer can call another subnetwork when it runs a certain subnetwork, and it can also call it recursively. The use of lexical expansion transfer networks makes it possible to interact with the syntactic processing phase of language understanding, and effectively resolves ambiguities in Chinese word splitting.

– Matrix constraint method

A syntactic constraint matrix and a semantic constraint matrix are created, in which the elements indicate whether the adjacency of a word with a certain lexical nature and another lexical nature conforms to the syntactic rules, and whether the adjacency of a word belonging to a semantic class and another word belonging to a semantic class conforms to the logic, respectively, and the machine uses them to constrain the result of word separation when slicing.

- Statistical model learning-based word separation method

The statistical model learning word division method is also known as the dictionary-free word division method. Words are stable combinations, so the more times adjacent words occur together in a context, the more likely they are to form a word. Thus the probability or frequency of word-adjacent occurrences gives a better indication of the confidence of forming a word. The word separation method of statistical model learning counts the frequency of each combination of words appearing next to each other in the training text and calculates the mutual occurrence information between them. The reciprocal information reflects the closeness of the bonding relationship between characters. When the closeness is higher than a certain threshold, it is considered that the word group may form a word.

The commonly used statistical models are N-gram, Hidden Markov Model (HMM), Maximum Entropy Model (ME), Conditional Random Fields (CRF), etc.

Such word separation algorithms in practical applications are generally combined with lexicon-based word separation methods to take advantage of both the speed and efficiency of matching word separation cuts, and the advantages of lexicon-free word separation combined with contextual recognition of raw words and automatic disambiguation.

- Deep learning neural network based word separation method

The method works by simulating the parallel, distributed processing and building numerical computational models of the human brain. It deposits the decentralized implicit method of word separation knowledge inside the neural network, modifies the internal weights through self-learning and training to achieve the correct word separation result, and finally gives the automatic word separation result of the neural network.

Common neural network models include LSTM, GRU and other neural network models.

7.3.2 Lexical Annotation Technology

The role of lexical annotation in the problem analysis module is to determine the lexical nature of the words in the text, i.e. to determine which lexical nature each word belongs to. The lexical nature of words is divided into five categories: verb, noun, adverb, adjective or other lexical nature. The lexical annotation process is essential in problem analysis and even in natural language processing research, both for English and Chinese. Because of the universality of lexical annotation, it plays a huge role in the whole linguistic research and has achieved excellent results in several fields, most notably in information retrieval and text classification.

Three popular algorithms are included in the lexical annotation approach.

- rule-based annotation algorithms, which themselves all contain manually annotated rule bases that consume significant manual costs.

- random annotation-based algorithms, which, to be implemented, must have a large amount of data as a training data set to obtain a model that is used to determine the likelihood that a word in the text is of which lexical nature, such as HMM-based annotation algorithms.
- Hybrid annotation algorithm, which combines the advantages of the first two algorithms to achieve the optimal comprehensive performance, is widely used in natural language processing, such as TBL annotation algorithm.

7.3.3 Deactivation Technology

Deactivated words are commonly understood as "words that are dummy words" or "words that are not valid for retrieval". The presence of deactivated words tends to affect the response rate of text-based question and answer robots and can take up a lot of storage space in the experimental machine. Therefore, in order to save storage space and improve search efficiency when retrieving answers to questions, these deactivated words are automatically eliminated so that they do not affect the efficiency of the system in answering questions.

Deactivated words are not the same as filter words that are often mentioned because the latter contain more types of words, such as yellow, political and other words that are often viewed as filter words and then processed with human settings for the filter words. Filter words are often coupled with human settings, where people process words that do not need to appear, while deactivated words do not require human intervention.

7.3.4 Feature Extraction Technology

The role of feature extraction is to transform the data in the dataset into matrix data that can be used directly by deep learning methods. In this process feature extraction is only responsible for transforming the data form, and the rest of the factors are not taken into account without the need to understand the availability of features. Further feature selection enables the selection of representative subsets of features in the transformed feature set that can represent useful information about the text.

Deep learning methods cannot directly process the original text data, which must be specifically processed and transformed, and the result of feature extraction is the data that generates the data that deep learning methods can process.

The data text provided by the dataset is represented in the form of text. To extract the information in the text, the data text needs to be pre-processed (word separation process) first, and the result of word separation is converted into a fixed-length vector through the representation of word vector, and the newly generated vector can be directly recognized and processed by the deep learning network at a later stage.

The specific steps of feature extraction are as follows.

- The words occurring in the original dataset are counted according to the statistical algorithm and constitute the initial lexical vector. The newly generated vector is composed of all words in the original dataset, and all words (assuming that deactivated words have been removed) can find their corresponding elements in the newly generated vector.
- All texts can be represented as vectors after the first step of processing. Each text can be represented as a dictionary vector of its own unique length, and the length of the dictionary vector varies if the text is different.
- Generally, we use the 0–1 representation to describe the text, if a word appears then the corresponding vector element is represented as 1, if it does not appear then the corresponding vector element is represented as 0.

Because feature extraction does not analyze the useless information in the text, it is converting all the text into lexicon vectors, so the dimensionality of its generated lexicon vectors is high, which is not conducive to direct computation. Therefore, the feature vectors involved in the computation in the later stage are the vectors after feature selection, and the feature selection reflects the role of dimensionality reduction in this link to avoid the problem of dimensional disaster in the computation.

7.3.5 *Problem Classification Technology*

The purpose of problem classification is to understand the user's intent when he or she asks a question by first classifying the question into different categories and then going deeper. The problem of problem classification is often viewed as how to solve a mapping function that maps a problem $x \in X$ into a certain category, as in Eq. (7.3).

$$f: X \rightarrow \{y_1, y_2 ..., y_n\} \tag{7.3}$$

where f is expressed as based on prior empirical knowledge.

y_i belongs to the set of category Y.
X is the set of questions.

In the problem analysis phase, problem classification serves two purposes. On the one hand, it reduces the candidate space of answers to a certain extent. On the other hand, the answer extraction strategy is determined by the category of the question, and the corresponding answer selection strategy set knowledge base is different for different categories of questions.

7.3.6 Answer Extraction Technology

Answer extraction, i.e. information extraction from structured, semi-structured, unstructured and other data with different structures, identifies, discovers and extracts information such as concepts, types, facts, relationships, rules, etc. that constitute the answer.

Structured information has a strong structure, often automatically generated by program control, and the object of information extraction is generally the content corresponding to certain fields; unstructured information has a strong syntax, such as news information in web information; semi-structured information is between the two, and its information content is ungrammatical and has a certain format, but there is no strict control. When semi-structured information and unstructured information are used for answer extraction, they can also be transformed into structured text, and the answer extraction of structured information is carried out.

7.4 Typical Applications of Internet-Based Text Questions and Answers Robots

The continuous development of computer information and Internet technology has led to the development of various online services in the direction of networking, intelligence and automation. The rapid development of mobile Internet and Internet e-commerce has given rise to the application of text question and answer (Q&A) robots in Internet e-commerce, mobile APP, and WeChat channels. Text Q&A robots interact with users through websites, APPs, WeChat and other channels to answer questions from users, which reduces the service costs of enterprises and institutions on the one hand, and optimizes the user experience on the other.

Currently, the most widely used Internet-based text Q&A robots with good results are the restricted domain text Q&A robots based on the FAQ knowledge base.

7.4.1 System Architecture of Text Questions and Answers Robots Based on FAQ Restricted Domain

The system architecture of text questions and Answers based on the FAQ restricted domain text Q&A robot system also contains question analysis, information retrieval, and answer extraction modules, while applied to the Internet environment, in order to meet the needs of massive user interaction and the needs of multiple interaction channels, while ensuring the high availability, high concurrency, scalability, and security of the robot system, the system adopts a distributed design and is divided from top to robottom into an access layer, an interaction layer, service layer, and data layer. The overall structure of the system is shown in Fig. 7.5.

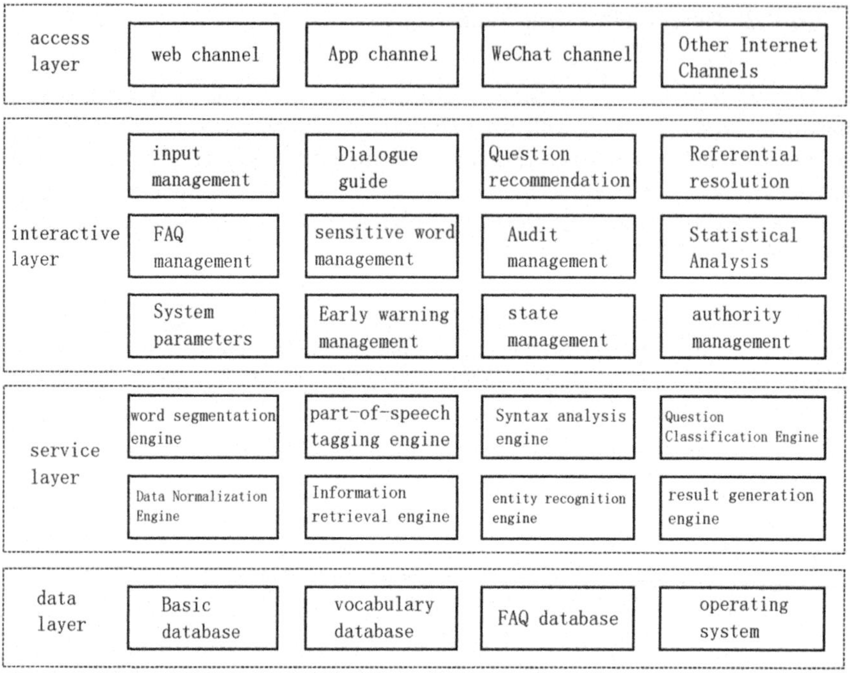

Fig. 7.5 Structure of FAQ-based restricted domain text-based Q&A robot system

The access layer is the interface docking and information distribution layer of the system, docking the network interfaces of web, APP, WeChat and other network channels, and then sending different front-end information to the application management for processing according to the rules. The access layer optimizes the management of information distribution of the system, and the access layer affects the scope and availability of application channels of the system.

The interaction layer is the application interaction management system of the system, which manages the information from the access layer in modules, including information management of user input content, sensitive word management, FAQ knowledge base management, knowledge audit management, knowledge status management, and parameter and permission management of the system. The interaction layer best reflects the functions of the system and the interaction experience of users.

The service layer is the technical core of the system. The engines corresponding to question analysis, information retrieval and answer extraction of the system are managed in the service layer, including the word separation engine, lexical annotation engine, question classification engine, data normalization engine, information retrieval engine, entity identification engine, result generation engine, etc. The service layer affects the accuracy and recall of the system, and is the key to distinguish different systems and determine the system performance index.

The data layer is the service platform related to data storage. It contains on the one hand the operating system data for managing the robot system as well as the file system, but also the FAQ database, the word list database of the system, and the basic database of the application system. The data layer affects the overall performance of the system.

7.4.2 System Functions of Text Questions and Answers Robots Based on FAQ Restricted Domain

The FAQ-based restricted-domain text-based Q&A robot system has been practically applied for a long time, and the system has achieved good overall performance and user experience. In addition to performing simple one-on-one questions and answers, the robot system also supports complex functions and application scenarios such as scenario-based Q&A, referential disambiguation, and associated business systems.

- User input information pre-processing

The pre-processing of user input information includes the filtering of invalid information and the normalization of user input information. At the same time, in order to improve the effect of user Q&A and the accuracy of the system, the system access layer matches the knowledge in the FAQ library according to the keywords for the incomplete information input by users, and carries out real-time dynamic prompting during the user's question input process to guide users to adopt the standard questions in the FAQ library for Q&A. In order to avoid the problem of lower accuracy rate caused by ambiguous word splitting and unregistered word problems.

- User problem identification

User question recognition is the basic function of traditional text Q&A robots, i.e., matching the knowledge corresponding to the user's question in the FAQ database. User question recognition is the core function of the system. Based on the final question information sent by the user, the system gets the normalized user questions through question division, lexical annotation, syntactic analysis, sentence type analysis, question classification, question formalization, and formal expansion; the conditions and requirements for information retrieval are obtained through the question processing results, and the information retrieval engine is invoked to retrieve all the information that meets the conditions from the FAQ library. Then the search results are initially filtered according to the question type; according to the results of information retrieval, the system sorts the information that has been initially filtered and processes the information that ranks first as the final answer back to the user.

- Scenario-based Q&A

Scenario-based Q&A is designed to cope with the problem of low accuracy of Q&A due to the large number of FAQ libraries. Scenario-based Q&A is a process in which users get different answers to the same questions they enter under different contextual conditions.

Scenario-based Q&A requires the construction of different Q&A scenarios in the FAQ library, and the contextual relationships of the Q&A scenarios are related in a tree structure. The elements of each Q&A scenario contain entry, process control and exit mechanisms.

- Entrance: The system carries out entrance control through semantic matching of utterances to distinguish whether different dialogues are ordinary Q&A or scenario-based Q&A. The essence of entrance control and identification is the process of user question identification.
- Process control: Process control, i.e., user dialogue management in the scenario, after the user's dialogue based on the above, the first question after entering the scenario Q&A is the entrance question, and the second and subsequent questions are semantically related to the above, such as the user can answer "yes", "right", "Shanghai household" and other positive, negative or explicit questions, and also ambiguous questions such as "okay", "average" and so on. The Q&A robot retrieves and matches different knowledge nodes of the tree knowledge in the FAQ database according to the user's contextual intent.
- Exit mechanism: The system matches the semantics of statements according to the user's contextual semantics and sets up a scenario Q&A exit mechanism. When the user's following reply statement does not match the knowledge of the target node in the scenario, the exit condition of the scenario is satisfied. At this time, the user's question is then retrieved and matched with other knowledge in the FAQ library.

- Related Questions Recommendation

The text Q&A robot system based on the restricted domain of FAQ has a better post-processing capability of Q&A. Based on the result of the answer generation process, if the weight of the knowledge with the largest weight satisfies the threshold condition, the answer corresponding to that knowledge is the final answer, and at the same time, the questions of other knowledge that satisfy another weight threshold condition and are ranked according to the weight are recommended.

Relevant question recommendations are another effective way to address the low resolution rate of the system.

- Knowledge Learning

Knowledge learning is a way to collect similar questions of the FAQ base or information not contained in the FAQ base. Knowledge learning is divided into two approaches: classification approach and clustering approach.

The classification approach is to divide each knowledge as a class by calculating the relationship between the information to be learned and the features or attributes of the known knowledge; the clustering approach is to calculate the Euclidean distance between the information to be learned and the centroid knowledge for the division.

7.4.3 System Features of Text Questions and Answers Robots Based on FAQ Restricted Domain

Text Q&A robot system based on FAQ restricted domain can be widely used in the browser-based Internet domain. Due to the wide application and technical maturity of WEB technology, text Q&A robots based on FAQ restricted domain has the advantages of high concurrency, high availability, high security, and high accuracy, while its answer range is limited due to the content limitation of the FAQ library.

- High concurrency: The system can be deployed in clusters and distributed search technology, which can guarantee the simultaneous dialogue interaction of a large number of users.
- High availability: Due to the distributed structure design, the system can perform parameter optimization, web-side code tuning, compression, caching, reverse proxy optimization, and operating system file handle number optimization at the front access layer to enable the system to support a high information throughput rate.
- System security: At the network level, the system delineates security areas, deploys firewall systems, deploys security audit systems, vulnerability scanning systems and network virus monitoring systems; at the system level, the system ensures the security of the system by means and methods such as host intrusion prevention, malicious code prevention and resource control.
- High accuracy: The system adopts different word separation methods in the process of Chinese word separation according to different and application fields; gradually enriches the word list in the application process; adopts user problem pre-processing and relevant problem recommendation in the application process to make up for the limitations of the algorithm model and ensure the high accuracy of the system.
- Limited scope of Q&A: Due to the limited scope of the number of information in the FAQ library, the system can only be applied to a specific field or a specific service scenario, and the Q&A effect is generally very poor for questions beyond this scope.

7.4.4 Application Areas of Text Questions and Answers Robots Based on FAQ Restricted Domain

Since the text Q&A robot system, which is based on FAQ restricted domain, has a high accuracy rate of Q&A in specific domains and scenarios, generally exceeding 80%, the robot is widely used in specific domains for specific user groups.

At present, the text Q&A robot system based on FAQ restricted domain has been widely used in the fields of government affairs consultation, hospital registration process Q&A, bank credit card common operation dialogue, internal human resources and financial system human–machine inquiries, etc. A large number of repetitive, high-frequency questions are answered by robots instead of humans, which not only significantly reduces the human resources cost in the communication process, but also improves the real time of related questions and answers, which has achieved good results.

Correction to: Special Robot Technology

Correction to:
T. Guo et al., *Special Robot Technology*, Advanced and Intelligent Manufacturing in China, https://doi.org/10.1007/978-981-99-0589-8

The original version of the book was inadvertently published with an incorrect Figure and Equation. The Figure and Equation corrections for chapters 3 and 5 have been corrected. The correction chapters and book have been updated with the changes.

The updated versions of these chapters can be found at
https://doi.org/10.1007/978-981-99-0589-8_3,
https://doi.org/10.1007/978-981-99-0589-8_5

T. Guo et al., *Special Robot Technology*, Advanced and Intelligent Manufacturing in China, https://doi.org/10.1007/978-981-99-0589-8_8

Bibliography

1. Zixing, Cai. 2015. *Robotics*, 3rd ed. Beijing: Tsinghua University Press.
2. Shuguang, Wang. 2013. *Mobile robot principle and design*. Beijing: People's Post and Telecommunications Publishing House.
3. Yi, Zhang. 2007. *Mobile robotics and its applications*. Beijing: Electronic Industry Press.
4. Huangxiang, Chen. 2012. *Intelligent robot*. Beijing: Chemical Industry Press.
5. Yaonan, Wang. 2004. *Robot intelligent control engineering*. Beijing: Science Press.
6. Shiqiang, Zhu. 2000. *Robotics and its applications*. Hangzhou: Zhejiang University Press.
7. Guofu, Gao. 2004. *Robot sensors and their applications*. Beijing: Chemical Industry Press.
8. Tongying, Guo. 2016. *Robotics and its intelligent control*, 2nd ed. Beijing: Chemical Industry Press.
9. Ruqing, Yang. 2000. *Intelligent control engineering*. Shanghai: Shanghai Jiaotong University Press.
10. Bin, Li. 2003. Research on snake robot and its application in disaster rescue. *Robotics and Applications* 3: 22–26.
11. Xiang, Chen. 2013. Rescue robots participate in Sichuan Ya'an earthquake rescue. *Robotics and Applications* 3: 46.
12. Xing, Liu. 2013. *Vision sensor-based coordination control of mobile robots on rugged ground*, 24–27. Harbin: Harbin Engineering University.
13. Lili, Zhou, Yan He, Xiaoying Tian, and Tao Wang. 2013. Static target targeting and tracking system for mobile robots on rugged ground. *Automation Technology and Applications* 32 (5): 4–8.
14. Jiajun, Gu. 2010. *Research on autonomous navigation of mobile robots on non-flat terrain*, 117–135. Shanghai: Shanghai Jiao Tong University.
15. Xin, Zhang, Xiaodong Yang, Lili Guo, and Shu Zhang. 2005. Study on leakage cables for wireless communication in closed or semi-closed space. *Journal of Harbin Engineering University* 05: 672–674.
16. Niku, Saeed B. 2018. Introduction to robotics: analysis, control and applications, 2nd edn [Sun Fuchun, Translation]. Beijing: Electronic Industry Press.
17. Jinxia, Yu, Zixing Cai, Xiaobing Zou, and Duan Tuohua. 2005. Research on the projection method of mobile robot trajectory under non-flat terrain. *Journal of Henan University of Science and Technology* 24 (3): 210–216.
18. Hongru, Tang, Aiguo Song, and Xiaobing Zhang. 2005. Research on autonomous stair climbing technology for mobile robots based on sensor information fusion. *Journal of Sensing Technology* 18 (4): 828–833.
19. T.Q. Li. 2008. *Research on autonomous stair climbing of robots based on multi-sensor fusion*, 44–49. Hefei: Hefei University of Technology.
20. X.D. Zhai, R. Liu, Q.F. Hong, and Yuan Qu. 2010. Analysis of the ladder climbing mechanism of an articulated tracked mobile robot. *Mechanics and Electronics* 1: 62–65.

T. Guo et al., *Special Robot Technology*, Advanced and Intelligent Manufacturing in China, https://doi.org/10.1007/978-981-99-0589-8

21. B.R. Hong. 2005. Research on autonomous charging of mobile robots in indoor environment. *Journal of Harbin Institute of Technology* 37 (7): 885–887.
22. Zhongmin, Wang. 2007. Current status and development trend of disaster search and rescue robot research. *Modern Electronics Technology* 17 (30): 152–155.
23. Wei, Zhao. 2013. *Research on robot positioning accuracy improvement technology based on laser tracking measurement*. Hangzhou: Zhejiang University.
24. Chengqing, Zong. 2013. *Statistical natural language processing*, 2nd ed., 57–67. Beijing: Tsinghua University Press.
25. J.G. Liu, Y.C. Wang, B. Li, and Shugen Ma. 2006. Simulation analysis of tipping stability of deformed robot. *Journal of Instrumentation* 18 (2): 409–415.
26. J.G. Liu, Y.C. Wang, B. Li, and Shugen Ma. 2006. Expression and counting of non-isomorphic configurations for modular deformable robots. *Journal of Mechanical Engineering* 42 (1): 98–105.
27. Youzheng, Wu., Jun Zhao, and Xiangyu Duan. 2013. A review of question-and-answer retrieval techniques and evaluation research. *Chinese Journal of Information* 19 (3): 11–13.
28. Yinfei, Huang, Fang Zheng, and Pengwei Yan. 2001. Design and implementation of EasyNav, a campus navigation system. *Chinese Journal of Informatics* 15 (4): 35–40.
29. Wang, Shuxi, Qun Liu, and Shuo Bai. 2013. An expert system for character relationship question and answer system. *Journal of Guangxi Normal University* 21 (1): 31–36.
30. J.T. Zhang, and Y.P. Du. 2013. Application of semantic chain-based retrieval in QA systems. *Computer Science* 40 (2): 256–260.
31. Tian Xia, Xiaozhong Fan, Lin Liu, et al. 2004. A natural language interface for Chinese based on ALICE. *Journal of Beijing University of Technology* 24 (10): 885–889.
32. S.F. Meng, and S. Wang. 2001. Design and implementation of Nchiql, a natural language query system for Chinese databases. *Computer Research and Development* 38 (9): 1080–1086.
33. Kun, Liu. 2016. *Research on unmanned ship path planning based on artificial potential field and ant colony algorithm*. Hainan: Hainan University.
34. Zhenjie, Geng. 2015. *Research on multi-objective path planning of ball picking robot based on improved ant colony algorithm*. Lanzhou: Lanzhou University of Technology.
35. W.R. Shi, Huang X.H, and W. Zhou. 2010. Path planning for mobile robots based on improved artificial potential field method. *Computer Applications* 08: 2021–2023.
36. Y. Wang, , H. Zhu, Y.S. Wang, et al. 2007. Current status of research on coal mine disaster relief robots and the technical problems that need to be focused on. *Coal Mine Machinery* 28 (4): 107–109.
37. J. Xu, Z. Guo, and T.H. Lee. 2014. Design and implementation of integral sliding-mode control on an underactuated two-wheeled mobile robot. *IEEE Transactions on Industrial Electronicso* 61 (7): 3671–3681.
38. Siciliano, Bruno, and Oussama Khatib. 2008. Springer Handbook of robotics. Springer Press
39. R.C. Luo, and C.C. Lai. 2014. Multisensor fusion-based concurrent environment mapping and moving object detection for intelligent service robotics. *IEEE Transactions ON Industrial Electronics* 61 (8): 4043–4051.
40. Asif, M., M.J. Khan, and N. Cai. 2014. Adaptive sliding mode dynamic controller with integrator in the loop for nonholonomic wheeled mobile robot trajectory tracking. *International Journal of Control* 87 (5): 964–975.
41. Y. Mao, and H. Zhang. 2014. Exponential stability and robust H-infinity control of a class of discrete-time switched non-linear systems with time-varying delays via T-S fuzzy model. *International Journal of Systems Science* 45 (5): 1112–1127.
42. Blazic, S. 2014. On periodic control laws for mobile robots. *IEEE Transactions on Industrial Electronics* 61 (7): 3660–3670.
43. Robin. 1996. Strategies for searching and area with semi-autonomous mobile robots. In *Proceedings of robotics for challenging environments*, 15–21.
44. Wang, Ahelong, and Hong Gu. 2007. A review of locomotion mechanisms of urban search and rescue robot. *Industrial Robot: An International Journal* 400–411.

45. Chien, Ting, Jr Guo, Kuo Su, and Sheng Shiau. 2007. Develop a multiple interface based fire fighting robot. In *Proceedings of international conference on mechatronics Kumamoto Japan*, 1–6.
46. Kruijff, Geert Jan M. 2012. Rescue robots at earthquake-hit Mirandola, Italy: a field report. In *IEEE international symposium on safety, security, and rescue robotics*, 1–8.
47. Zelinsky, Alexander. 2012. *Field and service robotics*, 79–85. Springer Publishing Company.
48. Cardeira, Carlos, and Jose Sada Costa (2005) A low cost mobile robot for engineering education. In *Industrial electronics society, 31st annual conference of IEEE*, 2162–2167.
49. Mirats Tur, Josep, and Carlos Pfeiffer. 2006. Mobile robot design in education. *IEEE Robotics & Automation Magazine* 69–75.
50. Sabe, Kohtaro. 2005. Development of entertainment robot and its future. *Symposium on VLSI Circuits Digest of Technical Papers* 1–5.
51. Paul, Hebert, and Bajracharya Max. 2015. Mobile manipulation and mobility as manipulation design and algorithms of RoboSimian. *Journal of Field robotics* 32: 255–274.
52. Satzinger. 2015. Tractable locomotion planning for RoboSimian. *The International Journal of Robotics Research* 19–25.
53. Jijkoun, V., and Rijke M. 2005.Retrieving answers from frequently asked questions pages on the web. In *Proceedings of NIKM*, 76–83
54. 2014. Reconfigurations for RoboSimian. In *ASME 2014 dynamic systems and control conference, American society of mechanical engineers,* 120–127.
55. Satzinger, Bajracharya. 2014. More solutions means more problems: resolving kinematic redundancy in robot locomotion on complex terrain. In *IEEE international conference on intelligent robots and systems,* 4861–4867.
56. Robin, Murphy. 2000. Marsupial and shape-shifting robots for urban search and rescue. In *IEEE international conference on intelligent systems,* 14–17.
57. Miyanaka, Hitoshi, Norihiko Wada, and Tetsushi Kamegawa. 2007. Development of an unit type robot "KOHGA2" with stuck avoidance capability. In *IEEE international conference on robotics and automation,* 3877–3882.
58. Folkesson, Christensen. 2008. SIFT based graphical SLAM on a Packrobot. *Springer Tracts in Advanced robotics* 42: 317–328.
59. Cheung, Grocholsky. 2008. UAV-UGV collaboration with a Packrobot UGV and Raven SUAV for pursuit and tracking of a dynamic target. *Unmanned Systems Technology X* 65–72.
60. Rudakevych, Pavlo. 2007. Integration of the Fido explosives detector onto the Packrobot EOD UGV. *ProcSpie* 61–65.
61. Eich, Markus, Felix Grimminger, and Frank Kirchner. 2008. A versatile stair-climbing robot for search and rescue applications. In *IEEE international workshop on safety, security and rescue robotics,* 35–40.
62. Tongying, Guo, Peng Liu, and Haichen Wang. 2014. Design and implementation on PC control interface of robot based on VxWorks operating system. In *International conference on precision mechanical instruments and measurement technology,* 1109–1112.
63. Hesheng, Wang, Maokui Jiang, and Weidong Chen. 2011. Visual servoing of robots with uncalibrated robot and camera parameters. *Mechatronics* 187–192.
64. Hesheng, Wang, Maokui Jiang, and Weidong Chen. 2010. Adaptive visual servoing with imperfect camera and robot parameters. In *International conference on intelligent for sustainable energy and environment*, 255–261.
65. Vincent, Isabelle, and Qiao Sun. 2012. A combined reactive and reinforcement learning controller for an autonomous tracked vehicle. *Robotics and Autonomous Systems.* 60: 599–608.
66. Anastasios, Moutikis. 2007. Autonomous stair climbing for tracked vehicles. *The International Journal of robotics Research* 60 (7): 737–758.
67. Matsuno, Tadokoro 2005. Rescue robots and systems in Japan. In *IEEE international conference on robotics and biomimetics*, 12–20.
68. Murphy, Casper. 2000. Mobility and sensing demands in USAR. In *IEEE international conference on session and rescue engineering*, 138–142.

69. Scholtz, Antonishek, and Young. 2005. A field study of two techniques for situation awareness for robot navigation in urban search and rescue. In *IEEE international workshop on robot and human interactive communication*, 131–136.
70. Minghui, Wang, and Shugen Ma. 2006. Motion planning for a reconfigurable robot to cross an obstacle. In *IEEE international conference on mechatronics and automation*, 1291–1296.
71. Changlong, Ye, Shugen Ma, Bin Li. 2006. Design and basic experiments of a shape-shifting mobile robot for urban search and rescue. In *IEEE international conference on intelligent robots and systems*, 3994–3999.
72. Minghui, Wang, and Shugen Ma. 2005. Task planning and behavior scheduling for a reconfigurable planetary robot system. In *IEEE International Conference on Mechatronics and Automation* 729–734.
73. Tonglin, Liu, Chengdong Wu, and Bin Li. 2010. Shape-shifting robot path planning method based on reconfiguration performance. In *IEEE international conference on intelligent robots and systems*, 4578–4583.
74. Bin, Li, Shugen Ma, Tonglin Liu, and Minghui Wang. 2010. Cooperative reconfiguration between two specific configurations for a shape-shifting robot. In *IEEE international workshop on safety security and rescue robotics*, 1–6.
75. Jinguo, Liu, Yuechao Wang, Bin Li, Shugen Ma, Jing Wang, and Huibin Cao. 2006. Transformation technique research of the improved link-type shape shifting modular robot. In *IEEE international conference on mechatronics and automation*, 295–300.
76. Minghui, Wang, Shugen Ma, and Bin Li. 2009. Reconfiguration of a group of weel-manipulator robots based on MSV and CSM. *IEEE Transactions on Mechatronics* 14 (2): 229–239.
77. Bin, Li, Jing Wang, Jinguo Liu, Yuechao Wang, and Shugen Ma. 2006. Study on a novel link-type shape shifting robot. In *The sixth world congress on intelligent control and automation*, 9012–9016.
78. Minghui, Wang, Shugen Ma, and Bin Li. 2006. Configuration analysis for reconfigurable modular planetary robots based on MSV and CSM. In *IEEE international conference on intelligent robots and system*, 3191–3196.
79. Tonglin, Liu, Chengdong Wu, Bin Li, and Jinguo Liu. 2010. A path planning method for a shape-shifting robot. In *The eighth World Congress on Intelligent Control and Automation*, 96–101.
80. Abacha, A.B., and P. Zweigenbaum. 2015. MEANS; a medical question-answering system combining NLP techniques and semantic web technologies. *Information Processing & Management* 51 (5): 570–594.
81. Atzori, M., C. Zaniolo. 2015. Expressivity and accuracy of by-example structured queries on wikipedia. In *2015 IEEE 24th international conference on enabling technologies: infrastructure for collaborative enterprises. IEEE computer society*, 239–244.
82. Fang, Y., and L. Si. 2015. Related entity finding by unified probabilistic models. *World Wide Web-intemet & Web Information Systems* 18 (3): 521–543.
83. S. Zhang, B. Wang, and J.F. Gareth. 2014. ICT-DCU question answering task at NTCIR-6. In *Proceedings of NTCIR-6 workshop meeting*, 15–18. Tokyo: National Institute of Informatics.
84. L.D. Wu, X.J. Huang, Y.Q. Zhou,et al. 2003. FDUQA on TREC2003 QA task. In *The twelfth text retrieval conference (TREC2003)*, 246–253. Maryland: NIST.
85. Pavlic, M., D. Han, and A. Jakupovic. 2015. Question answering with a conceptual framework for knowledge-based system development a node of knowledge. *Expert Systems with Applications* 42(12):5264–5286.
86. Park, S., H. Shim, S. Han, et al. 2015. Multi-source hybrid question answering system. In *Natural language dialog systems and intelligent assistants*, 241–245. Springer International Publishing.
87. MInock, and Michael. 2005. Where are the killer applications' of restricted domain question answering?. In *Proceedings of the IJCAI workshop on knowledge reasoning in question answering*, vol. 4.
88. Stutzle, T., and H. Hoos. 2000. Max-min ant system. *Journal of Future Generation Computer Systems* 16 (9): 889–914.

89. Ioannidis, K., GCh. Sirakoulis, and I. Andreadis. 2011. Cellular ants: a method to create collision free trajectories for a cooperative robot team. *Robotics and Autonomous Systems* 59 (2): 113–127.
90. Chandra Mohan, B., and R. Baskaran. 2012. A survey: ant colony optimization based recent research and implementation on several engineering domains. *Expert Systems with Applications* 39 (4): 4618–4627.

The manufacturer's authorised representative in the EU is Springer Nature Customer Service Centre GmbH, Europaplatz 3, 69115 Heidelberg, Germany. If you have any concerns regarding our products, please contact ProductSafety@springernature.com

Printed and bound by CPI Group (UK) Ltd, Croydon, CR0 4YY
01/10/2026
02225872-0004